Organic Farming for Sustainable Agriculture

Organic Farming for Sustainable Agriculture

Edited by Cruz Hawkins

SYRAWOOD
PUBLISHING HOUSE

New York

Published by Syrawood Publishing House,
750 Third Avenue, 9th Floor,
New York, NY 10017, USA
www.syrawoodpublishinghouse.com

Organic Farming for Sustainable Agriculture
Edited by Cruz Hawkins

© 2019 Syrawood Publishing House

International Standard Book Number: 978-1-68286-779-2 (Hardback)

Cataloging-in-Publication Data

Organic farming for sustainable agriculture / edited by Cruz Hawkins.
    p. cm.
Includes bibliographical references and index.
ISBN 978-1-68286-779-2
1. Organic farming. 2. Sustainable agriculture. 3. Agriculture. I. Hawkins, Cruz.
S605.5 .O74 2019
631.584--dc23

TABLE OF CONTENTS

PREFACE

Sustainability in agriculture is achieved with due consideration to ecological processes such as nutrient cycling, nitrogen fixation, soil regeneration and implementation of practices that are environment-friendly. Organic farming is a sustainable alternative to traditional farming practices. It relies on using organic fertilizers like bone meal, compost and green manure for agricultural productivity. It also lays emphasis on the implementation of techniques like crop rotation and companion planting. This method of farming opposes the use of synthetic fertilizers, pesticides, genetically modified organisms, growth regulators and hormones in the farming process. Naturally occurring pesticides such as rotenone and pyrethrin are used in organic farming. Weed management in organic farms involves mulching, tillage, flame weeding, mowing and cutting. This book is a compilation of chapters that discuss the most vital concepts and emerging trends in the field of organic farming. While understanding the long-term perspectives of the topics, the book makes an effort in highlighting their impact as a modern tool for the growth of the discipline. This book will serve as a reference to agriculturists, horticulturists, agro-economists, agroecologists and other experts involved in this area of study.

This book is the end result of constructive efforts and intensive research done by experts in this field. The aim of this book is to enlighten the readers with recent information in this area of research. The information provided in this profound book would serve as a valuable reference to students and researchers in this field.

At the end, I would like to thank all the authors for devoting their precious time and providing their valuable contribution to this book. I would also like to express my gratitude to my fellow colleagues who encouraged me throughout the process.

Editor

Conventional and Organic Farming — Does Organic Farming Benefit Plant Composition, Phenolic Diversity and Antioxidant Properties?

Alfredo Aires

Additional information is available at the end of the chapter

Abstract

The growing demanding from consumers for healthier foods, produced using environmentally friendly farming practices has resulted in the rapid expansion of organic farming. There are numerous studies about the importance of organic farming but the majority of the results are sometimes contradictory, inconsistent and show no clear link between organic farming practices and enhancement of the nutritional quality of plant-derived foods. As such, ongoing research into the effects of organic farming and cultivation practices in comparison with intensive farming, is very important. The objective of this chapter is to discuss the most recent data and variation in the responses of plants to farming regimes in order to better understand the relationship between agricultural practices and high levels of valuable compounds (glucosinolates, phenolics, minerals, vitamins, antioxidants), as well as low levels of undesirable components such as nitrates, nitrites and microorganisms.

Keywords: Organic farming, conventional faming, nutrient diversity, phytochemicals, quality, safety

1. Introduction

Research studies continue to show that the desire of consumers to be able to purchase healthier fruits and vegetables, produced by a more sustainable and environmental friendly agricultural system, is increasing day-by-day. The majority of these studies attempt to show how safe and nutritious organic foods are for humans [1] and animals [2]. According to European regulations

[3] organic farming is defined as an overall system of farm management and food production that combines the best environmental practices, high levels of biodiversity, the preservation of natural resources, the application of high animal welfare standards and utilises production methods in line with the preference of consumers for products produced using natural substances and processes. The aim of an organic farming system is to provide to the consumer with fresh, tasty and natural food, while respecting natural systems and the environment. To achieve this, several principles and rules are followed in order to minimize human impact on the environment, while at the same time ensuring the agricultural system operates as naturally as possible [4]. Several different approaches are employed, but all of them are guided by strict rules [3] aimed at protecting the integrity of the environment, plants, animals and biodiversity.

A fundamental aim of organic farming is the provision of healthy, high quality plant and animal-derived foods. The concept of food quality can be defined in many different ways. Often, the quality of food is based on visual characters such as shape, size and colour, but can also be described as containing fewer pesticides, or more nutrients, or even containing specific functional properties due to elevated levels of phytochemicals [1, 5]. Thus, there is no one sole concept of quality. Nonetheless, countless studies of quality always refer to at least one or more of the following criteria: (i) food safety (absence of undesirable components like nitrites and pathogenic microorganisms); (ii) primary nutrients (minerals and vitamins, for example); (iii) secondary metabolites and phytochemicals that are closely associated with the beneficial health properties of plant and animal-derived foods; and (iv) observed health effects. However, research studies using these criteria vary widely, with investigative topics ranging from the taste of the food to how the food in question benefits health. Despite this diversity, the link between organic products and their nutritional, functional, and biological values is far from being fully understood. Therefore, in this chapter, we discuss recent advances in organic farming, particularly its differences from conventional farming, highlighting the differences in vitamins, minerals, phytochemicals, antioxidant activity and sensorial properties.

2. Factors and constraints affecting crop and plant-derived food composition

Growing crops in any part of the world is affected by many variables, including environmental, agronomical, social and economic factors, among others. These factors can affect not which particular type of agricultural system is employed, or which type of crop produced, but also and more importantly, the quality of the crop. Both conventional and organic farming systems are always heavily influenced by such factors. These factors can be grouped into 4 main types (Figure 1): a) socio-economic; b) pre-harvest; c) harvest; and d) post-harvest.

A recent study [6] showed that the choice between an organic or conventional farming system is primarily dependent on socio-economic factors, secondarily dependent on social aspects and then all of the remaining factors follow on. In fact, when farmers implement any production system or crop, their first question is: How profitable is it to produce? The answer will depend on the choices the farmer makes about what crops to grow, where to grow them, and

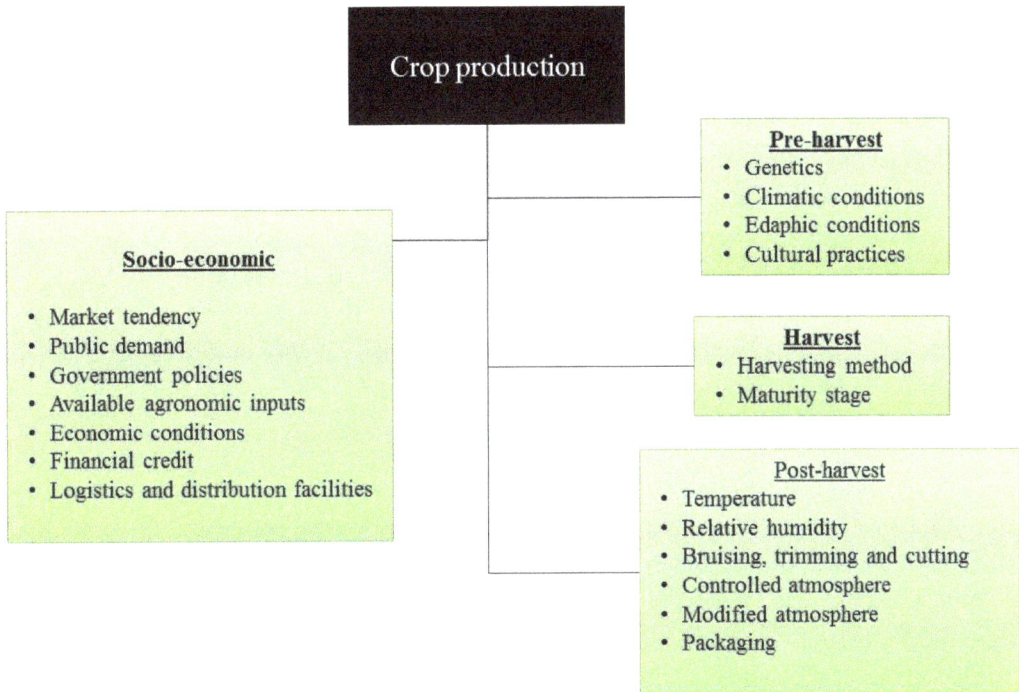

Crop production

Pre-harvest
- Genetics
- Climatic conditions
- Edaphic conditions
- Cultural practices

Harvest
- Harvesting method
- Maturity stage

Post-harvest
- Temperature
- Relative humidity
- Bruising, trimming and cutting
- Controlled atmosphere
- Modified atmosphere
- Packaging

Socio-economic
- Market tendency
- Public demand
- Government policies
- Available agronomic inputs
- Economic conditions
- Financial credit
- Logistics and distribution facilities

Figure 1. Constraint factors of any crop yield and production.

what technologies he uses. In addition, farmers tend to follow the system producing a higher financial income, lower financial risks, lower labour requirements, and if possible, the greatest pleasure [7]. The ability to obtain credit will also influence the choice of crops, farming systems and technologies [8]. The level of technical and scientific knowledge of production will also affect a farmer's propensity to choose a particular crop or production system [9, 10]. Moreover, the capital requirement for any crop development is always present, but can vary seasonally and is often far higher during harvesting than at other times during the production period. Any financial or labour constraint can negatively affect negatively the farmer's productivity and, therefore, income [11].

Another social aspect of decision to farm organically or conventionally is public demand [12]. If a farmer wants to succeed, then there must be a demand for their products, to generate an income, otherwise the farmer will switch to another, more profitable crop, whether it is organic or not.

Production is also affected by pre-harvest factors. In general, these factors include all physical factors, such as genetics, geology, soil and climatic conditions and cultural practices [13, 14, 15]. In other words, after a specific crop has been chosen, its success will depend on the outcome of the complex interaction between numerous elements such as the biology of the plant, interaction between plant and soil, crop management techniques, mineral and organic nutrition, chemical or biochemical treatments, and the watering regime employed, among other factors. Climatic parameters such temperature, humidity, altitude, rainfall and wind, are all fundamental factors affecting the variation of plant and crop success [16, 17] and thus their nutritional quality as food. Temperatures can limit the growth of crops; water is a key factor in plant growth with different crops requiring water at different times; altitude primarily affects the average temperatures and consequently the type of farming; wind can have a destructive effect on crops physically, as well as increasing the dryness of soils, reducing moisture and increasing the potential for soil erosion. The soil type will influence crop cultivation because different crops prefer different soils, e.g., clay soils with their high levels of water retention are widely used to produce rice, as rice requires a lot of of water to grow successfully [18, 19], whilst sandy soils are more suited to roots, tubers and vegetables, due to their need for better drainage, which is a requirement for good development of their roots [20]. Thus, selecting the right crop for the given specific conditions is fundamental to increasing yield and quality.

Another set of factors are relate to the harvest period. It is widely accepted that stage of maturity at harvest can have a critical influence on the nutritional content of the crop. Zaro et al. [21], observed marked changes in the level of bioactive compounds present (anthocyanins, carotenoids, ascorbic acid, phenolics) and in antioxidant activity of purple eggplants at the fruiting stage. They found a decrease of such compounds and beneficial properties when plants were harvested at earlier stages (I and II). The same tendency was recently observed [22] in carrots, where a relatively high amount of falcarindiol, an important antioxidant compound, was present during very early harvest (i.e. 103 to 104 days after sowing) compared with a later harvest (i.e. 117 to 118 days after sowing). The same trend was also recently noted [23] for anthocyanin content in blueberries when harvested earlier, but not when harvested at full maturity. Thus, correct choice of harvesting time is crucial in preserving the quality of fresh produce during storage. This way, it is possible to provide the consumer with high quality fresh food products.

After harvesting, several factors (identified here as post-harvest factors) can interfere with the quality of fruit and vegetables. Among them are temperature regime of storage, relative humidity of storage, type of atmosphere used if any, and packaging [24, 25]. Temperature management during shelf-life is one of the most important means of preserving the quality of fresh roots, fruits and vegetables. After harvest, any delay in cooling, or choosing the wrong temperature regime, can result in losses in nutritional quality, flavour, taste and saleability. Tano et al. [26], found that the quality of mushrooms, tomatoes and cabbages stored under a fluctuating temperature regime was severely affected by extensive browning, loss of firmness, increased weight loss, increased level of ethanol in plant tissues, and fungal infections due to physiological damage and excessive condensation, when compared with products stored at a constant temperature. Similar observations were recently made [27] for mandarins, when low storage temperatures (2, 5 and 8 $^{\circ}$C) resulted in a loss of orange peel colour, volatile compounds, and flavour. Thus, storage temperature is a fundamental factor affecting nutrients, colour and flavour [27]. In addition, particular attention should be paid post-harvest procedures such as cleaning, bruising, trimming and cutting, which may also affects the quality of products if they are conducted in inappropriate conditions or improperly performed [28]. Thus, the quality and stability of plant-derived food products will be strongly dependent on the interaction of several different factors and, therefore, an understanding of the physiological and biochemical process in plants and foods during the period of shelf-life, is crucial to maximising their nutritional quality and bioactive composition, and thereby their properties beneficial to health.

3. Conventional versus organic

Organic farming has increased in popularity in recent decades due to the public's perception that health problems may arise from the consumption of plant-derived foods produced under intensive farming practices. This growing concern lead to a considerable number of studies into the effect of organic production on nutrients (mineral, vitamins) and phytochemicals such as polyphenols, antioxidant vitamins (A, C, E), glucosinolates, carotenoids and isoflavones, among others. Although a large number of studies about the differences between plants produced under conventional and organic farming systems is now available, most of the studies present contradictory facts, inconsistent results and the differences are often reported as negligible. Consequently, it is important to study the variation in nutritional quality and safety of plant-derived food produced under both organic and conventional farming methods. In the following paragraphs we discuss recent findings about the effect of the two different agricultural systems on the variation in nutrients and phytochemicals in plant-derived food, focusing on the major differences already discovered.

3.1. Variations in vitamin, mineral, amino-acid and nitrate content

The nutritional value of food is essentially a function of its vitamin and mineral content, particularly those related to important beneficial functions in animals and humans [29]. Essential minerals required in the human diet include, among others, phosphorus (P), potassium (K), calcium (Ca), magnesium (Mg), iron (Fe), sulphur (S), boron (B), chromium (Cr), cobalt (Co), copper (Cu), iodine (I), manganese (Mn), molybdenum (Mo), selenium (Se), tin (Sn), and zinc (Zn) [30, 31] and the essential vitamins include mainly A, B (all vitamins of the B complex), C, E and K [30]. Compared with conventional farming, organic production relies on sustainable management practices, which include crop rotations, cover cropping, nutrient recycling, integrated pest management, and use of organic fertilisation [32], among other practices. All these practices, according to the majority of consumers have indeed had a positive impact on food quality, enhancing the levels of beneficial minerals and vitamins [33]. However, from a scientific point of view, the question of whether organic plant-derived foods are more nutritious than conventional ones remains.

Conventional farming usually relies on massive doses of readily soluble forms of mineral fertilisers (mainly in N, P, K form), whilst organic farming relies on the incorporation of organic material into the soil, normally through the use of animal manure as fertiliser [34]. Composted manure is the most commonly used fertiliser in organic farming [35] and thus the general consumer perception is that organic foods are better because they are produced using natural and safe agronomical inputs [33], and thus they are more nutritious.

Throughout the past 15 years, several comparative studies have demonstrated significant differences in the content of vitamins, minerals and free amino-acids (Table 1). However, several authors claim that no major or significant differences are found in mineral and vitamin content in fruits and vegetables produced under organic or conventional farming systems, and several others report that for some specific nutrients, conventionally grown plant-derived foods usually contain higher average levels (Table 1).

Products tested	Nutrients analysed	Key-results	Reference
Lettuce, spinach, carrots, potato and cabbage	Iron, Mg, and P	**Higher** in **organics**	[36]
Chinese mustard, Chinese kale, lettuce, spinach	Vitamin C, β-carotene and riboflavin	**Higher** in **organics**	[37]
Wheat	Minerals (N, K, Mg, Ca, S, Fe,)	**Similar** in **both**	[38]
Red potatoes	Minerals (K, Mg, P, S and Cu)	**Higher** in **organics**	[39]
Wheat	Essential Amino acids	**Lower** in **organics**	[40]
Wheat	Minerals (P, K, Ca, Zn, Mo, Co)	**Similar** in **both**	[40]
Kiwi fruits	Minerals (N, P, K, S, B, Ca, Mg)	**Higher** in **organics**	[41]
Tomato	Vitamin C	**Lower** in **organics**	[42]
Broccoli	Vitamin C	**Similar** in **both**	[43]
Spinach	Nitrate	**Lower** in **organics**	[44]
Strawberry	Vitamin C	**Similar** content	[45]
Broadbean, bean, lettuce, pepper, watermelon.	Nitrates	**Lower** in **organics**	[46]
Acerola	Vitamin C and carotenoids	**Higher** in **organics**	[47]
Strawberries	Vitamin C and carotenoids	**Similar** in **both**	[47]
Cauliflower	Vitamin C	**Higher** in **organics**, but only when higher organic fertiliser levels were applied	[48]
Potatoes	Essential amino acids	**Higher** in **organics**	[49]
Strawberries	Ascorbic acid	**Higher** in **organics**	[50]
Cauliflower	Soluble solids, nitrates, P and K	**Similar** in **both**	[51]
Green pepper	Weight, firmness, thickness, N and P	**Lower** in **organics**	[52]
Tomatoes	Vitamin C	**Higher** in **organics**	[53]
Apple	Aromatic volatiles, organic acids and sugars	**Higher** in **organics**	[54]

Table 1. Differences in the content of nutrients in organic and conventional fruit and vegetables

Some research studies have claimed that organic amendments can have a positive effect on the content of antioxidant vitamins such as vitamin C [47], but others claims that the effect is negative [42], whilst others still, claim no significant difference [43, 45, 55]. Thus, there is a discrepancy in the results, and external factors such as crop variety, crop location, climate and growing conditions [56] can all exert an effect. Moreover, it is unlikely that mineral fertilisers or manure alone can affect the nutritional content of fruits and vegetables. Nonetheless, the majority of authors seems to agree that an organic production system is friendlier than an intensive or conventional farming system and the choice of organic system as an alternative to conventional practice can be justified by its lower environmental impact [57].

Another important issue related to the nutritional quality and safety of organic food is nitrate content, particularly in fresh vegetables. Nitrates are a natural consequence of the mechanism by which plants absorb the element nitrogen, in the form of NO3-, from fertilisers or organic material [58]. Although nitrate is an important component of plants, it has the potential to accumulate in tissues, particularly in green leafy vegetables [59] and thus, nitrate from fertilizers could accumulate in vegetables on a large scale. The danger of this, lies in the fact that nitrates can be reduced to nitrites, which can react with amines and amides to produce "N-nitroso" compounds, responsible for gastric cancer [60]. In order to maximize the health benefits from eating vegetables, measures should be taken to reduce levels of nitrates and nitrites [59]. This is particularly true in organic farming due to the large quantities of manure used as natural fertiliser, which is sometimes reported as having the potential to elevate levels of nitrates and nitrites up to, or above, maximum residue levels (MRLs), which is dangerous. However, some studies report that manure fertilisers have no significant effect on nitrate levels because organic products should always contain fewer nitrates than their counterparts produced by conventional methods, due to their lower concentration of nitrogen-based fertilisers [61, 62]. Furthermore, several other authors have reported that nitrate content is more closely related to genotype, soil conditions, growth conditions (i.e., nitrate uptake, nitrate reductase activity, and growth rate), storage and transport conditions, than to mineral or organic amendments [63]. More recently [64] it was shown that that nitrate accumulation in vegetables is more closely related to the quality of water and water accumulation in vegetable tissues. Thus, the results available until now from various different studies are sometimes contradictory and doubts still remain. Nonetheless, based on the fact that organic farming enhances specific nutrients and is less aggressive to the environment, it is more beneficial than conventional farming, which is seen as more aggressive to the environment, fauna and flora, and ultimately, to animals and humans.

3.2. Influence on bioactive compounds and functional properties of foods

3.2.1. Glucosinolates, phenolics, carotenoids and pigments

Recent scientific advances in plant-derived foods studies have mainly focused on the potential health effects of phytochemicals in plant foods. Phytochemicals, also known as bioactive compounds, are naturally occurring substances in plants, functioning mainly as secondary metabolites [65]. Their distribution in plants is considered to be the result of the natural adaptation of plants to environmental stress, pathogen infection, insects and other pests [66]. According Harbone [66], phytochemicals can be divided into different classes: phenolics (e.g. phenolic acids, flavonoids, anthocyanin), terpenoids (e.g., carotenoids, xanthophylls and other pigments), alkaloids (e.g., indole compounds), and sulphur-containing compounds (e.g., glucosinolates). Table 2 gives a brief summary of phytochemicals commonly found in fruits and vegetables, and the potential health benefits associated with them. To date, studies have shown that phytochemicals can have a protective effect on human health (Table 2 and Table 3), including mopping-up free radicals, reduction of oxidative stress, inhibition of cell prolif-eration, induction of cell differentiation, inhibition of oncogene expression, suppression of gene expression in carcinogenic processes, modulation of detoxification enzymes, stimulation of the immune system, regulation of hormone metabolism, and antibacterial and antiviral effects [67]. Strong associations have been also found between disease risk reduction and consumption of foods with a high content of glucosinolates (anti-cancer), tocopherols (cardi-ovascular), phenolics and carotenoids (eye-health) [68].

Phytochemicals		Example of food sources	Proposed health benefits found in literature
Class	Example		
Phenolic acids	Gallic acid, caffeic acid,	Tea, kiwi fruit, strawberries, pineapple, coffee	Antioxidant and anti-inflammatory
Flavonols	Quercetin	Red and yellow onions, tea, wine, apples, cranberries, beans	Antioxidant, anti-inflammatory, enzyme inhibitor and immune modulation
Flavanols	Catechins	Chocolate, tea, grapes, wine, apples, cocoa, black-eyed peas	Antioxidant, anti-hypertensive, anti-inflammatory, anti-proliferative, anti-thrombogenic, and lipid lowering effects
Flavones	Apigenin	Chamomile, celery, parsley	Lowers high blood pressure, antioxidant and anti-inflammatory
Anthocyanins	Cyanindin	Blackberry, blueberries, red wine, strawberries	Improvement of vision, and neuroprotective effects
Isoflavones	Genistein	Soy, alfalfa sprouts, red clover, chickpeas other legumes	Reduction in blood pressure, antioxidant activity
Lignans	Secoisolariciresinol	Linseed, sunflower seeds, sesame seeds, pumpkin seeds	Improves glucose control, prevents pre-cancerous cellular changes, decreases the incidence of several chronic diseases
Stilbenes	Resveratrol	Grape skins and seeds, wine, nuts and peanuts	Antioxidant, anti-inflammatory, protects the body against nitric oxide, keeps the blood vessels optimally dilated

Phytochemicals		Example of food sources	Proposed health benefits found in literature
Class	Example		
Carotenoids	Lycopene, beta-carotene and other types of carotenes	Carrots, spinach, tomato and several other types of fruits and vegetables	Neutralisation of free radicals that cause cell damage
Monterpenes	Limonene	Citrus oils, cherries, spearmint, garlic, maize, rosemary, basil	Antioxidant, anti-inflammatory, anti-cancer, helps with weight management ("fat cleanser") and helps clear cholesterol
Diterpenes	Gingkolides	Gingko biloba	Protects neurons against Abeta1-42-induced synapse damage and cognitive loss
Triterpenes	Ginsenosides	Ginseng	Boosts the immune system and may lower blood sugar levels
Phytosterols	Sitosterol	Sunflower oil, avocados, rice bran, peanuts, soybeans	Inhibits 5-alpha reductase in prostate tissue
Alkaloids	Capsaicin	Chili pepper	Reduces the expression of proteins that control growth genes that cause malignant cells to grow
Glucosinolates, isothiocyanates	Sulforaphane, allyl-isothiocyanate, phenethyl-isothiocyanate,	Broccoli, mustard, cress, cabbages and all Cruciferae family plants	Neutralisation of free radicals that causes cell damage. Protection against some cancers
Indoles	Alliin, allicin	Onions, garlic, leeks	Antimicrobial agents and decreases LDL cholesterol

Table 2. Examples of some important phytochemicals commonly found in foods

- antioxidant activity
- neutralises free radicals and reduces oxidative stress
- inhibition of cell proliferation
- induction of cell differentiation
- inhibition of oncogene expression
- inhibition of tumour gene expression
- induction of cell cycle arrest
- induction of apoptosis
- inhibition of signal transduction pathways

- phase II enzyme
- glutathione peroxidase (GPX)
- catalase
- superoxide dismutase (SOD)
- enzyme inhibition
- phase I enzyme (block activation of carcinogens)
- cyclooxygenase-2 (COX-2)
- inducible nitric oxide synthase (iNOS)
- xanthine oxide

- enhancement of immune functions and surveillance
- anti-angiogenesis
- inhibition of cell adhesion and invasion
- inhibition of nitrosation and nitration
- prevention of DNA binding
- antibacterial and antiviral effects

• enzyme induction and enhancing
detoxification

---

[1] Adapted from Liu and Finley [67].

Table 3. Proposed health protective mechanisms of dietary phytochemicals[1]

Glucosinolates are sulphur-containing compounds mainly present in the Cruciferae family. When consumed, they are hydrolysed via myrosinase (EC 3.2.1.147, thioglucoside glucohydrolase) into isothiocyanates (ITCs) and other derivative products [69], that up-regulate genes associated with carcinogen detoxification cellular mechanisms [70]. Clinical studies have shown that the products of glucosinolate hydrolysis can reduce the incidence of certain forms of cancer [71].

Other compounds such as carotenoids lutein, β-carotene and tocopherols in addition to their role as vitamins, are also powerful antioxidants [72]. Tocopherols and carotenoids have been associated with the decrease of certain forms of cancer [73] and with a reduction in risk of cardiovascular diseases [74], whilst lutein protects against the development of cataracts and age-related macular degeneration [75], even if according Trumbo and Ellwood [76] there is no credible scientific evidence to support a health claim that lutein or zeaxanthin intake can reduce the risk of age-related macular degeneration or cataracts.

Phenolic compounds are a large group of secondary metabolites, categorised according to their chemical structure, into different classes, with phenolic acids, flavonoids, stilbenes and lignans being the most relevant ones [77]. They all have in common the presence of labile hydrogen able to neutralise or mop-up free radicals, and as such they are recognised as powerful antioxidants. Fruits and vegetables are the richest potential sources of these substances [78].

As mentioned above, the diversity of the chemical composition of plants, and thus by extension of phytochemicals is determined by a number of factors, including genotype, ontogeny, growth conditions, management practices and the environment. Thus, it might be expected that differences caused by organic vs. conventional growing practices may cause associated differences in phytochemical levels and diversity. Increasing organic food consumption is partially as a result of consumer perception that organic foods are healthier, but do organic foods actually contain more phytochemicals than conventional foods? Are the levels of phytochemicals in organic production relevant? Is the diversity of phytochemicals in foods affected by agronomical practices?

Table 4 summarises some of the results from different studies conducted over the last 15 years into the difference in phytochemical content in fruits and vegetables produced under organic and conventional farming practices. This is not an exhaustive list, but unsurprisingly several different conclusions are drawn. Recent studies [79, 80, 53] have indicated that organic produce contains higher concentrations of certain phytochemicals associated with health, than those produced under conventional farming systems. In addition, some studies [81, 82] reinforce this idea, stating that the abiotic and biotic stress induced by organic farming practices seems to overcome the variability among samples and consequently, the use of organic practices may be a means of increasing the levels of phytochemicals. However, according a recent observation [83] there is little evidence for any differences in the health benefits of organic and conventional produce. The differences often found may in fact be due to cultivar genotype influence and climatic variation rather than agricultural practices. The same observations was made by Oh et al. [84] and Lv et al. [85].

Crops & products	Bioactive substances	Key-results	Reference
Apple	Polyphenols	**Higher** in **organic** production	[86]
Chinese cabbage	Flavonoids	**Higher** in **organic** production	[87]
Spinach	Flavonoids	**Higher** in **organic** production	[87]
Green pepper	Flavonoids	**Higher** in **organic** production	[87]
Pear	Polyphenols	**Higher** in **organic** production	[88]
Yellow plum	Quercetin	**Lower** in **organic** production	[89]
Apple	Anthocyanins	**Higher** in **organic** production	[90]
Tomato	Lycopene	**Similar** content in both systems	[91]
Broccoli	Total glucosinolates	**Lower** in **organic** production	[92]
Strawberry	Polyphenols	**Similar** content in both systems	[93]
Tomato	Carotenes	**Higher** in **organic** production	[94]
Tomato	Polyphenols	**Lower** in **organic** production	[95]
Blueberry	Polyphenols	**Higher** in **organic** production	[96]
Tomato	β-carotene	**Higher** in **organic** production	[42]
Tomato	Lycopene	**Lower** in **organic** production	[42]
Carrot	Carotenoids	**Similar** content in both systems	[97]
Egg-plant pulp	Phenolics	**Similar** content in both systems	[98]
Cauliflower	Glucosinolates	**Similar** content in both systems	[48]
Strawberry	Anthocyanins	**Higher** in **organic** production	[79]
Soybeans	Isoflavones	**Lower** in **organic** production	[99]
Broccoli and collard greens	Glucosinolates	**Higher** in **organic** production	[100]
Watercress	Glucosinolates	**Lower** in **organic** production	[100]
Broccoli	Glucosinolates	**Lower** in **organic** production	[80]
Tomato	Polyphenols and lycopene	**Higher** in **organic** production	[53]
Pepper	Higher	**Lower** in **organic** production	[101]
Broccoli	Polyphenols	**Similar** content in both systems	[102]
Broccoli	Glucosinolates	**Higher** in **organic** production	[102]

Table 4. Summary of studies comparing phytochemical contents in fruits and vegetables from organic and conventional production

These authors stated that the most important factor affecting the phytochemical composition of plants is the interaction between genotype, environment and agronomical practices. Therefore, it is crucial to select the optimal environment conditions, genotype and best agronomical practices, in order to maximise the levels of a components beneficial to health.

In order to accurately evaluate the differences between organic and conventional farming systems, all the factors affecting quality of produce must be controlled, which is a major limitation of some studies through their poor experimental design. So, an accurate evaluation of all these aspects should be made over a substantial period of time (more than one year at least) in order to assess the eventual changes related to the year, seasonal effect, genotype or agronomical practices employed. A multi-year sampling study to evaluate farming systems with the necessary consistency to draw valid conclusions, is a minimum requirement [103].

3.2.2. Antioxidant activity

Closely linked to phytochemical content is the variation in antioxidants. Antioxidants, by definition, are any substance that reduce or inhibit oxidation or other reactions caused by oxygen and peroxides and free radicals, and which protect the body from the deleterious effects of free radicals [104]. Well-known antioxidants includes enzymes, vitamins (C and E), carotenes, polyphenols and others capable of counteracting the damaging effects of oxidation. They are important, because to date, epidemiological studies have shown their preventive effect against several infectious processes such as cancer, and neurodegenerative and cardio-vascular diseases [105, 106, 62, 81]. As with primary nutrients and phytochemicals, the effect of organic farming practices on the antioxidant properties of plant-derived foods is contro-versial. It is common to find an association between organic farming practices and an increase in antioxidant content, and the converse is also true (Table 5).

Wang [81] found that organic practices result in an increase antioxidant activity in blueberries (measured by the ORAC) due to the increase of phenolic acids and anthocyanin content when compared with a conventional system, whilst Garuso and Nardini [107], didn't find any substantial difference in antioxidant activity in wines produced under organic and conven-tional farming practices. Similar observations were made by Unal et al. [108] for Brassicacea vegetables. They didn't detect any significant difference in antioxidant activity in brassicas produced under organic and conventional practices. However, Stracke et al. [97], when comparing the organic and conventional cultivation of apples over three years, observed that organic apples presented on average 15% higher antioxidant content, as determined by FRAP, TEAC and ORAC than conventionally produced fruits, but these authors also observed that inter-annual climatic variations were more critical to the antioxidant capacity than the type of farming. Despite these inconsistencies, the majority of authors seem to agree that the type of farming system may affect the phytochemical composition and thus by extension the amount of antioxidant activity. Since organic farming does not provide as much nitrogen as conven-tional fertilizers [56], as well as causing more stress to the plants (Straus et al., 2012)[109] than conventional farming, it has the potential to influence the synthesis of antioxidants, increasing their levels and thus increasing antioxidant activity, as recently reported [110]. Therefore, at least theoretically, it can be concluded that organic farming has a tendency to produce foods with more nutritional value, based on their enhanced antioxidant content and activity.

Crops & products	Antioxidant activity in organic compared to conventional counterpart	Reference
Blueberries	Higher in organics	[81]
Apples	Similar in both	[111]
Fruits and vegetables	Similar in both and no consistent trends were found	[112]
Tomato	Higher in organic	[113]
Grapes and wines	Higher in organics	[114]
Lettuce	Similar in both	[115]
Tomato	Higher in organics	[110]
Tomato	Higher in organics	[116]
Brassicas	Lower in organics	[108]
Oranges	Similar in both	[117]

Table 5. Some examples of studies comparing antioxidant activity of fruits and vegetables produced under organic and conventional farming practices

3.3. Consumers' sensory expectations and preferences related to variability of antioxidant activity and phytochemical content of organic foods

There is common belief that organic food is healthier and safer than conventional food. According to the vast amount of literature already published, some of which is reported in this chapter, organic food is free of chemical residues, contain fewer nitrates and more antioxidants. In respect of product quality, surveys in the last 10 years [118, 119, 120, 121, 122, 123] indicate that consumers consider organic foods to be more beneficial for human health than their conventional counterparts, even if those studies often assume a lack of knowledge on behalf of the consumers of the aims and production practices of organic farming. Moreover, consumers often buy organic foods based on an emotional view, such as a desire to preserve traditional products and processes [124]. According to a survey conducted in Turkey in 2012 [120] consumers indicated 4 main reasons to buy organic foods: they are healthier, they have higher quality, the price is normally acceptable, and the food is microbiologically safe. As Monk et al. reported in 2012 [125], for the majority of consumers, the idea of enhanced nutrition, being free from chemicals, and a better taste, are the major advantages of organic foods. Consumers often think that organic food is better because it tastes better, but apart from physical and sensorial qualities, the understanding of nutritional quality by consumers seems to be a question of the ability to find credible information [118], which they often can't. A recent survey [126] showed that 78% of consumers when questioned about the quality of labelling information, responded that they didn't believe that all food labelled 'organic' was, in fact, organic, and neither did they totally believe in their healthier effects. Often, consumers purchased organic food due to personal morals or beliefs such as: 'I feel obliged to buy organic food to protect my health' and 'I feel obliged to buy organic food to protect the health of my family' [126]. The same authors observed that consumers repeatedly reported that they experience difficulty in getting more knowledge about a product's properties, certification bodies, and labels etc... Nonetheless, nowadays consumers tend to be more conscious and more aware about the positive effects of organic foods on health and the environment [127], and as a result are buying more organic foods.

4. Conclusions

Since the 1980s, organic farming has been increasing due to growing demand from consumers for high quality foods, with lower pesticide residues, less synthetic fertilisers and produced using environmentally friendly practices. Presumably, animal and plant derived foods have fewer chemical residues and veterinary drugs in them when compared with conventional ones. The growing perception from consumers that organic foods are healthier and safer, has to the rapid growth of this type of production seen over the last 20 years. Although the beneficial properties of these foods for human health have not been unequivocally proven, the accumulation of nutritional metabolites in organic cultivation has been well documented. Recent studies have shown that organic foods are, from a nutritional point of view, at least similar to conventional ones, if not slightly better. Also, recent epidemiological studies advocate that under organic farming practices, plants can accumulate nutrients and phytochemicals, enhancing their biological value and thus increasing the nutritional quality of foods. Moreover, the growing evidence of lower pesticide exposure to consumers of organic foods, is one of the main reasons for converting to organic farming. Although more and more well-documented studies are still required to improve our understanding of which factors contribute to differences between organic and conventional farming practices, the most recent findings provide evidence-based knowledge that organic farming is a sustainable way of producing healthier and safer plant-derived foods.

Acknowledgements

The author acknowledges the financial support provided by the Portuguese Foundation for Science and Technology (FCT) (Alfredo Aires-SFRH/BPD/65029/2009) under the project UID/AGR/04033/2013.

Author details

Alfredo Aires

Address all correspondence to: alfredoa@utad.pt

Centre for the Research and Technology of Agro-Environment and Biological Sciences (CITAB), Universidade de Trás-os-Montes e Alto Douro, UTAD, Quinta de Prados, Portugal

References

[1] Huber M, Rembiałkowska E, Średnicka D, Bügel S, van de Vijver LPL. Organic food and impact on human health: Assessing the status quo and prospects of research. NJAS - Wageningen Journal of Life Sciences. 2011;58:103-109.http://dx.doi.org/10.1016/j.njas.2011.01.004

[2] Velimirov A, Huber M, Lauridsen C, Rembiałkowska E, Seidel K, Bügel S. Feeding trials in organic food quality and health research. Journal of the Science of Food and Agriculture. 2010;90:175-182. http://dx.doi.org/10.1002/jsfa.3805

[3] 189/1 RENEOJotEUL. Council Regulation (EC) No 834/2007 of 28 June 2007 on organic production and labelling of organic products and repealing EN Official Journal of the European Union L 189/1. 2007. http://eur-lex.europa.eu/legal-content/EN/TXT/PDF/?uri=CELEX:32007R0834&from=EN

[4] IFOAM. Definition of Organic Agriculture. 2009, http://www.ifoam.bio/en/organic-landmarks/definition-organic-agriculture

[5] Renaud ENC, Bueren ETLv, Myers JR, Paulo MJo, Eeuwijk FAv, Zhu N, et al. Variation in Broccoli Cultivar Phytochemical Content under Organic and Conventional Management Systems: Implications in Breeding for Nutrition. Plos One. 2014. http://dx.doi.org/10.1371/journal.pone.0095683

[6] Bowman MS, Zilberman D. Economic Factors Affecting Diversified Farming Systems. Ecology and Society. 2013;18:33. http://dx.doi.org/10.5751/ES-05574-180133

[7] Stoorvogel JJ, Antle JM, Crissman CC. Trade-off analysis in the Northern Andes to study the dynamics in agricultural land use. Journal of Environmental Management. 2004;72:23-33.http://dx.doi.org/10.1016/j.jenvman.2004.03.012

[8] Knowler D, Bradshaw B. Farmers' adoption of conservation agriculture: A review and synthesis of recent research. Food Policy. 2007;32:25-48.http://dx.doi.org/10.1016/j.foodpol.2006.01.003

[9] Chavas J-P, Kim K. Economies of diversification: A generalization and decomposition of economies of scope. International Journal of Production Economics. 2010;126:229-235.http://dx.doi.org/10.1016/j.ijpe.2010.03.010

[10] Sautter JA, Czap NV, Kruse C, Lynne GD. Farmers' Decisions Regarding Carbon Sequestration: A Metaeconomic View. Society & Natural Resources. 2010;24:133-147.10.1080/08941920903012502

[11] Bosi S, Magris F. Endogenous business cycles: Capital–labor substitution and liquidity constraint. Journal of Economic Dynamics and Control. 2002;26:1901-1926. http://dx.doi.org/10.1016/S0165-1889(01)00013-6

[12] Smith EG, Jill Clapperton M, Blackshaw RE. Profitability and risk of organic production systems in the northern Great Plains. Renewable Agriculture and Food Systems. 2004;19:152-158,

[13] Tiwari U, Cummins E. Factors Influencing β-Glucan Levels and Molecular Weight in Cereal-Based Products. Cereal Chemistry Journal. 2009;86:290-301.10.1094/CCHEM-86-3-0290

[14] Søltoft M, Nielsen J, Holst Laursen K, Husted S, Halekoh U, Knuthsen P. Effects of Organic and Conventional Growth Systems on the Content of Flavonoids in Onions and Phenolic Acids in Carrots and Potatoes. Journal of Agricultural and Food Chemistry. 2010;58:10323-10329.10.1021/jf101091c

[15] Tiwari U, Cummins E. Factors influencing levels of phytochemicals in selected fruit and vegetables during pre- and post-harvest food processing operations. Food Research International. 2013;50:497-506.http://dx.doi.org/10.1016/j.foodres.2011.09.007

[16] Krishnan P, Ramakrishnan B, Reddy KR, Reddy VR. Chapter three - High-Temperature Effects on Rice Growth, Yield, and Grain Quality. In: Donald LS, editor. Advances in Agronomy. Volume 111: Academic Press; 2011. p. 87-206.

[17] Oloyede FM, Adebooye OC, Obuotor EM. Planting date and fertilizer affect antioxidants in pumpkin fruit. Scientia Horticulturae. 2014;168:46-50.http://dx.doi.org/10.1016/j.scienta.2014.01.012

[18] Bhattacharyya R, Prakash V, Kundu S, Srivastva AK, Gupta HS. Soil aggregation and organic matter in a sandy clay loam soil of the Indian Himalayas under different tillage and crop regimes. Agriculture, Ecosystems & Environment. 2009;132:126-134.http://dx.doi.org/10.1016/j.agee.2009.03.007

[19] Mekuria W, Getnet K, Noble A, Hoanh CT, McCartney M, Langan S. Economic valuation of organic and clay-based soil amendments in small-scale agriculture in Lao PDR. Field Crops Research. 2013;149:379-389.http://dx.doi.org/10.1016/j.fcr.2013.05.026

[20] Zotarelli L, Scholberg JM, Dukes MD, Muñoz-Carpena R, Icerman J. Tomato yield, biomass accumulation, root distribution and irrigation water use efficiency on a sandy soil, as affected by nitrogen rate and irrigation scheduling. Agricultural Water Management. 2009;96:23-34.http://dx.doi.org/10.1016/j.agwat.2008.06.007

[21] Zaro MJ, Keunchkarian S, Chaves AR, Vicente AR, Concellón A. Changes in bioactive compounds and response to postharvest storage conditions in purple eggplants as affected by fruit developmental stage. Postharvest Biology and Technology. 2014;96:110-117.http://dx.doi.org/10.1016/j.postharvbio.2014.05.012

[22] Kjellenberg L, Johansson E, Gustavsson K-E, Olsson ME. Polyacetylenes in fresh and stored carrots (Daucus carota): relations to root morphology and sugar content. Journal of the Science of Food and Agriculture. 2012;92:1748-1754.10.1002/jsfa.5541

[23] Eichholz I, Huyskens-Keil S, Rohn S. Chapter 21 - Blueberry Phenolic Compounds: Fruit Maturation, Ripening and Post-Harvest Effects. In: Preedy V, editor. Processing and Impact on Active Components in Food,http://dx.doi.org/10.1016/B978-0-12-404699-3.00021-4. San Diego: Academic Press; 2015. p. 173-180.

[24] Caleb OJ, Mahajan PV, Manley M, Opara UL. Evaluation of parameters affecting modified atmosphere packaging engineering design for pomegranate arils. International Journal of Food Science & Technology. 2013;48:2315-2323.http://doi.org/10.1111/ijfs.12220

[25] O'Grady L, Sigge G, Caleb OJ, Opara UL. Effects of storage temperature and duration on chemical properties, proximate composition and selected bioactive components of pomegranate (Punica granatum L.) arils. LWT - Food Science and Technology. 2014;57:508-515.http://dx.doi.org/10.1016/j.lwt.2014.02.030

[26] Tano K, Oulé MK, Doyon G, Lencki RW, Arul J. Comparative evaluation of the effect of storage temperature fluctuation on modified atmosphere packages of selected fruit and vegetables. Postharvest Biology and Technology. 2007;46:212-221.http://dx.doi.org/10.1016/j.postharvbio.2007.05.008

[27] Tietel Z, Lewinsohn E, Fallik E, Porat R. Importance of storage temperatures in maintaining flavor and quality of mandarins. Postharvest Biology and Technology. 2012;64:175-182.http://dx.doi.org/10.1016/j.postharvbio.2011.07.009

[28] Watada AE, Ko NP, Minott DA. Factors affecting quality of fresh-cut horticultural products. Postharvest Biology and Technology. 1996;9:115-125.http://dx.doi.org/10.1016/S0925-5214(96)00041-5

[29] Carneiro G, Laferrère B, Zanella MT. Vitamin and mineral deficiency and glucose metabolism – A review. e-SPEN Journal. 2013;8:e73-e79.http://dx.doi.org/10.1016/j.clnme.2013.03.003

[30] FAO. Vitamin and mineral requirements in human nutrition: report of a joint FAO/WHO expert consultation, Bangkok, Thailand, Second edition. 2001, http://www.fao.org/3/a-y2809e.pdf

[31] Campbell I. Macronutrients, minerals, vitamins and energy. Anaesthesia & Intensive Care Medicine. 2014;15:344-349. http://dx.doi.org/10.1016/j.mpaic.2014.04.003

[32] Wachter JM, Reganold JP. Organic Agricultural Production: Plants. In: Alfen NKV, editor. Encyclopedia of Agriculture and Food Systems;http://dx.doi.org/10.1016/B978-0-444-52512-3.00159-5. Oxford: Academic Press; 2014. p. 265-286.

[33] Falguera V, Aliguer N, Falguera M. An integrated approach to current trends in food consumption: Moving toward functional and organic products? Food Control. 2012;26:274-281.http://dx.doi.org/10.1016/j.foodcont.2012.01.051

[34] EU, Scientific Committee for Food. Opinion on nitrate and nitrite (expressed on 22 September 1995), Annex 4 to document III/56/95, CS/CNTM/NO3/20-FINAL.1995

http://ec.europa.eu/food/safety/docs/labelling_nutrition-special_groups_food-children-scf_reports_38_en.pdf

[35] Moral R, Paredes C, Bustamante MA, Marhuenda-Egea F, Bernal MP. Utilisation of manure composts by high-value crops: Safety and environmental challenges. Bioresource Technology. 2009;100:5454-5460.http://dx.doi.org/10.1016/j.biortech.2008.12.007

[36] Worthington V. Nutritional Quality of Organic Versus Conventional Fruits, Vegetables, and Grains. The Journal of Alternative and Complementary Medicine. 2001;7:161-173.10.1089/107555301750164244

[37] Ismail A FChnompmmnap. Determination of vitamin C, β-carotene and riboflavin contents in five green vegetables organically and conventionally grown.. Malaysian Journal of Nutrition. 2003;9:31-39, http://nutriweb.org.my/publications/mjn009_1/mjn9n1_art4.pdf

[38] Ryan MH, Derrick JW, Dann PR. Grain mineral concentrations and yield of wheat grown under organic and conventional management. Journal of the Science of Food and Agriculture. 2004;84:207-216.10.1002/jsfa.1634

[39] Wszelaki AL, Delwiche JF, Walker SD, Liggett RE, Scheerens JC, Kleinhenz MD. Sensory quality and mineral and glycoalkaloid concentrations in organically and conventionally grown redskin potatoes (Solanum tuberosum). Journal of the Science of Food and Agriculture. 2005;85:720-726.10.1002/jsfa.2051

[40] Mäder P, Hahn D, Dubois D, Gunst L, Alföldi T, Bergmann H, et al. Wheat quality in organic and conventional farming: results of a 21 year field experiment. Journal of the Science of Food and Agriculture. 2007;87:1826-1835.10.1002/jsfa.2866

[41] Amodio ML, Colelli G, Hasey JK, Kader AA. A comparative study of composition and postharvest performance of organically and conventionally grown kiwifruits. Journal of the Science of Food and Agriculture. 2007;87:1228-1236.10.1002/jsfa.2820

[42] Rossi F, Godani F, Bertuzzi T, Trevisan M, Ferrari F, Gatti S. Health-promoting substances and heavy metal content in tomatoes grown with different farming techniques. European Journal of Nutrition. 2008;47:266-272.10.1007/s00394-008-0721-z

[43] Wunderlich SM, Feldman C, Kane S, Hazhin T. Nutritional quality of organic, conventional, and seasonally grown broccoli using vitamin C as a marker. International Journal of Food Sciences and Nutrition. 2008;59:34-45.10.1080/09637480701453637

[44] Citak S, Sonmez S. Effects of conventional and organic fertilization on spinach (Spinacea oleracea L.) growth, yield, vitamin C and nitrate concentration during two successive seasons. Scientia Horticulturae. 2010;126:415-420.http://dx.doi.org/10.1016/j.scienta.2010.08.010

[45] Kahu K, Jänes H, Luik A, Klaas L. Yield and fruit quality of organically cultivated blackcurrant cultivars. Acta Agriculturae Scandinavica, Section B — Soil & Plant Science. 2008;59:63-69.10.1080/09064710701865139

[46] Herencia JF, García-Galavís PA, Dorado JAR, Maqueda C. Comparison of nutritional quality of the crops grown in an organic and conventional fertilized soil. Scientia Horticulturae. 2011;129:882-888.http://dx.doi.org/10.1016/j.scienta.2011.04.008

[47] Cardoso PC, Tomazini APB, Stringheta PC, Ribeiro SMR, Pinheiro-Sant'Ana HM. Vitamin C and carotenoids in organic and conventional fruits grown in Brazil. Food Chemistry. 2011;126:411-416.http://dx.doi.org/10.1016/j.foodchem.2010.10.109

[48] Picchi V, Migliori C, Lo Scalzo R, Campanelli G, Ferrari V, Di Cesare LF. Phytochemical content in organic and conventionally grown Italian cauliflower. Food Chemistry. 2012;130:501-509.http://dx.doi.org/10.1016/j.foodchem.2011.07.036

[49] Carillo P, Cacace D, De Pascale S, Rapacciuolo M, Fuggi A. Organic vs. traditional potato powder. Food Chemistry. 2012;133:1264-1273.http://dx.doi.org/10.1016/j.foodchem.2011.08.088

[50] Crecente-Campo J, Nunes-Damaceno M, Romero-Rodríguez MA, Vázquez-Odériz ML. Color, anthocyanin pigment, ascorbic acid and total phenolic compound determination in organic versus conventional strawberries (Fragaria × ananassa Duch, cv Selva). Journal of Food Composition and Analysis. 2012;28:23-30.http://dx.doi.org/10.1016/j.jfca.2012.07.004

[51] Maggio A, De Pascale S, Paradiso R, Barbieri G. Quality and nutritional value of vegetables from organic and conventional farming. Scientia Horticulturae. 2013;164:532-539.http://dx.doi.org/10.1016/j.scienta.2013.10.005

[52] López A, Fenoll J, Hellín P, Flores P. Physical characteristics and mineral composition of two pepper cultivars under organic, conventional and soilless cultivation. Scientia Horticulturae. 2013;150:259-266.http://dx.doi.org/10.1016/j.scienta.2012.11.020

[53] Vinha AF, Barreira SVP, Costa ASG, Alves RC, Oliveira MBPP. Organic versus conventional tomatoes: Influence on physicochemical parameters, bioactive compounds and sensorial attributes. Food and Chemical Toxicology. 2014;67:139-144.http://dx.doi.org/10.1016/j.fct.2014.02.018

[54] Raffo A, Baiamonte I, Bucci R, D'Aloise A, Kelderer M, Matteazzi A, et al. Effects of different organic and conventional fertilisers on flavour related quality attributes of cv. Golden Delicious apples. LWT - Food Science and Technology. 2014;59:964-972.http://dx.doi.org/10.1016/j.lwt.2014.06.045

[55] Kang Y, Khan S, Ma X. Climate change impacts on crop yield, crop water productivity and food security – A review. Progress in Natural Science. 2009;19:1665-1674.http://dx.doi.org/10.1016/j.pnsc.2009.08.001

[56] Dangour AD, Dodhia SK, Hayter A, Allen E, Lock K, Uauy R. Nutritional quality of organic foods: a systematic review. The American Journal of Clinical Nutrition. 2009;10.3945/ajcn.2009.28041.10.3945/ajcn.2009.28041

[57] Conti S, Villari G, Faugno S, Melchionna G, Somma S, Caruso G. Effects of organic vs. conventional farming system on yield and quality of strawberry grown as an annual or biennial crop in southern Italy. Scientia Horticulturae. 2014;180:63-71.http://dx.doi.org/10.1016/j.scienta.2014.10.015

[58] Gangolli SD, van den Brandt PA, Feron VJ, Janzowsky C, Koeman JH, Speijers GJA, et al. Nitrate, nitrite and N-nitroso compounds. European Journal of Pharmacology: Environmental Toxicology and Pharmacology. 1994;292:1-38.http://dx.doi.org/10.1016/0926-6917(94)90022-1

[59] Correia M, Barroso Â, Barroso MF, Soares D, Oliveira MBPP, Delerue-Matos C. Contribution of different vegetable types to exogenous nitrate and nitrite exposure. Food Chemistry. 2010;120:960-966.http://dx.doi.org/10.1016/j.foodchem.2009.11.030

[60] Savino F, Maccario S, Guidi C, Castagno E, Farinasso D, Cresi F, et al. Methemoglobinemia Caused by the Ingestion of Courgette Soup Given in Order to Resolve Constipation in Two Formula-Fed Infants. Annals of Nutrition and Metabolism. 2006;50:368-371, http://www.karger.com/DOI/10.1159/000094301

[61] Guadagnin SG, Rath S, Reyes FGR. Evaluation of the nitrate content in leaf vegetables produced through different agricultural systems. Food Additives & Contaminants. 2005;22:1203-1208.10.1080/02652030500239649

[62] González-Gallego J, García-Mediavilla MV, Sánchez-Campos S, Tuñón MJ. Fruit polyphenols, immunity and inflammation. British Journal of Nutrition. 2010;104:S15-S27.doi:10.1017/S0007114510003910

[63] Burns IG, Zhang K, Turner MK, Meacham M, Al-Redhiman K, Lynn J, et al. Screening for genotype and environment effects on nitrate accumulation in 24 species of young lettuce. Journal of the Science of Food and Agriculture. 2011;91:553-562.10.1002/jsfa.4220

[64] Burns I, Durnford J, Lynn J, McClement S, Hand P, Pink D. The influence of genetic variation and nitrogen source on nitrate accumulation and iso-osmotic regulation by lettuce. Plant and Soil. 2012;352:321-339.10.1007/s11104-011-0999-0

[65] Björkman M, Klingen I, Birch ANE, Bones AM, Bruce TJA, Johansen TJ, et al. Phytochemicals of Brassicaceae in plant protection and human health – Influences of climate, environment and agronomic practice. Phytochemistry. 2011;72:538-556.http://dx.doi.org/10.1016/j.phytochem.2011.01.014

[66] Harborne JB. Recent advances in chemical ecology. Natural Product Reports. 1989;6:85-109.10.1039/NP9890600085

[67] Liu RH, Finley J. Potential Cell Culture Models for Antioxidant Research. Journal of Agricultural and Food Chemistry. 2005;53:4311-4314.10.1021/jf058070i

[68] Higdon JV, Delage B, Williams DE, Dashwood RH. Cruciferous vegetables and human cancer risk: epidemiologic evidence and mechanistic basis. Pharmacological Research. 2007;55:224-236.http://dx.doi.org/10.1016/j.phrs.2007.01.009

[69] Fenwick GR, Heaney RK, Mullin WJ, VanEtten CH. Glucosinolates and their breakdown products in food and food plants. C R C Critical Reviews in Food Science and Nutrition. 1983;18:123-201.10.1080/10408398209527361

[70] Melchini A, Costa C, Traka M, Miceli N, Mithen R, De Pasquale R, et al. Erucin, a new promising cancer chemopreventive agent from rocket salads, shows anti-proliferative activity on human lung carcinoma A549 cells. Food and Chemical Toxicology. 2009;47:1430-1436.http://dx.doi.org/10.1016/j.fct.2009.03.024

[71] Zhang Y, Munday R, Jobson HE, Munday CM, Lister C, Wilson P, et al. Induction of GST and NQO1 in Cultured Bladder Cells and in the Urinary Bladders of Rats by an Extract of Broccoli (Brassica oleracea italica) Sprouts. Journal of Agricultural and Food Chemistry. 2006;54:9370-9376.10.1021/jf062109h

[72] Edge R, McGarvey DJ, Truscott TG. The carotenoids as anti-oxidants — a review. Journal of Photochemistry and Photobiology B: Biology. 1997;41:189-200.http://dx.doi.org/10.1016/S1011-1344(97)00092-4

[73] Wright ME, Virtamo J, Hartman AM, Pietinen P, Edwards BK, Taylor PR, et al. Effects of α-tocopherol and β-carotene supplementation on upper aerodigestive tract cancers in a large, randomized controlled trial. Cancer. 2007;109:891-898.10.1002/cncr.22482

[74] Tapiero H, Townsend DM, Tew KD. The role of carotenoids in the prevention of human pathologies. Biomedicine & Pharmacotherapy. 2004;58:100-110.http://dx.doi.org/10.1016/j.biopha.2003.12.006

[75] Moeller SM, Voland R, Tinker L, et al. ASsociations between age-related nuclear cataract and lutein and zeaxanthin in the diet and serum in the carotenoids in the age-related eye disease study (careds), an ancillary study of the women's health initiative. Archives of Ophthalmology. 2008;126:354-364.10.1001/archopht.126.3.354

[76] Trumbo PR, Ellwood KC. Lutein and zeaxanthin intakes and risk of age-related macular degeneration and cataracts: an evaluation using the Food and Drug Administration's evidence-based review system for health claims. The American Journal of Clinical Nutrition. 2006;84:971-974, http://ajcn.nutrition.org/content/84/5/971.abstract

[77] Spencer JPE, Abd El Mohsen MM, Minihane A-M, Mathers JC. Biomarkers of the intake of dietary polyphenols: strengths, limitations and application in nutrition research. British Journal of Nutrition. 2008;99:12-22. http://dx.doi.org/10.1017/S0007114507798938

[78] Kondratyuk TP, Pezzuto JM. Natural Product Polyphenols of Relevance to Human Health. Pharmaceutical Biology. 2004;42:46-63.doi:10.3109/13880200490893519

[79] Fernandes VC, Domingues VF, de Freitas V, Delerue-Matos C, Mateus N. Strawberries from integrated pest management and organic farming: Phenolic composition and antioxidant properties. Food Chemistry. 2012;134:1926-1931.http://dx.doi.org/10.1016/j.foodchem.2012.03.130

[80] Vicas S, Teusdea A, Carbunar M, Socaci S, Socaciu C. Glucosinolates Profile and Antioxidant Capacity of Romanian Brassica Vegetables Obtained by Organic and Conventional Agricultural Practices. Plant Foods for Human Nutrition. 2013;68:313-321.10.1007/s11130-013-0367-8

[81] Wang SY, Millner P. Effect of Different Cultural Systems on Antioxidant Capacity, Phenolic Content, and Fruit Quality of Strawberries (Fragaria × aranassa Duch.). Journal of Agricultural and Food Chemistry. 2009;57:9651-9657.10.1021/jf9020575

[82] García-Mier L, Guevara-González R, Mondragón-Olguín V, del Rocío Verduzco-Cuellar B, Torres-Pacheco I. Agriculture and Bioactives: Achieving Both Crop Yield and Phytochemicals. International Journal of Molecular Sciences. 2013;14:4203, http://www.mdpi.com/1422-0067/14/2/4203

[83] Smith-Spangler C, Brandeau ML, Hunter GE, Bavinger JC, Pearson M, Eschbach PJ, et al. Are Organic Foods Safer or Healthier Than Conventional Alternatives?A Systematic Review. Annals of Internal Medicine. 2012;157:348-366.10.7326/0003-4819-157-5-201209040-00007

[84] Oh M-M, Carey EE, Rajashekar CB. Environmental stresses induce health-promoting phytochemicals in lettuce. Plant Physiology and Biochemistry. 2009;47:578-583.http://dx.doi.org/10.1016/j.plaphy.2009.02.008

[85] Lv J, Lu Y, Niu Y, Whent M, Ramadan MF, Costa J, et al. Effect of genotype, environment, and their interaction on phytochemical compositions and antioxidant properties of soft winter wheat flour. Food Chemistry. 2013;138:454-462.http://dx.doi.org/10.1016/j.foodchem.2012.10.069

[86] Weibel FP TD, Haseli A, Graf U. Sensory and health related quality of organic apples: a comparative field study over three years using conventional and holistic methods to assess fruit quality. 1th International Conference on Cultivation Technique and Phyotpathological Problems in Organic Fruit Growing; LVWO: Weinsberg, Germany. 2004, 8 pp. http://orgprints.org/14536/

[87] Ren H, Endo H, Hayashi T. Antioxidative and antimutagenic activities and polyphenol content of pesticide-free and organically cultivated green vegetables using water-soluble chitosan as a soil modifier and leaf surface spray. Journal of the Science of Food and Agriculture. 2001;81:1426-1432.10.1002/jsfa.955

[88] Carbonaro M, Mattera M, Nicoli S, Bergamo P, Cappelloni M. Modulation of Antioxidant Compounds in Organic vs Conventional Fruit (Peach, Prunus persica L., and

Pear, Pyrus communis L.). Journal of Agricultural and Food Chemistry. 2002;50:5458-5462.10.1021/jf0202584

[89] Lombardi-Boccia G, Lucarini M, Lanzi S, Aguzzi A, Cappelloni M. Nutrients and Antioxidant Molecules in Yellow Plums (Prunus domestica L.) from Conventional and Organic Productions: A Comparative Study. Journal of Agricultural and Food Chemistry. 2004;52:90-94.10.1021/jf0344690

[90] Asami DK, Hong Y-J, Barrett DM, Mitchell AE. Comparison of the Total Phenolic and Ascorbic Acid Content of Freeze-Dried and Air-Dried Marionberry, Strawberry, and Corn Grown Using Conventional, Organic, and Sustainable Agricultural Practices. Journal of Agricultural and Food Chemistry. 2003;51:1237-1241.10.1021/jf020635c

[91] Caris-Veyrat C, Amiot M-J, Tyssandier V, Grasselly D, Buret M, Mikolajczak M, et al. Influence of Organic versus Conventional Agricultural Practice on the Antioxidant Microconstituent Content of Tomatoes and Derived Purees; Consequences on Antioxidant Plasma Status in Humans. Journal of Agricultural and Food Chemistry. 2004;52:6503-6509.10.1021/jf0346861

[92] Robbins RJ, Keck A-S, Banuelos G, Finley JW. Cultivation Conditions and Selenium Fertilization Alter the Phenolic Profile, Glucosinolate, and Sulforaphane Content of Broccoli. Journal of Medicinal Food. 2005;8:204-214.10.1089/jmf.2005.8.204

[93] Anttonen MJ, Hoppula KI, Nestby R, Verheul MJ, Karjalainen RO. Influence of Fertilization, Mulch Color, Early Forcing, Fruit Order, Planting Date, Shading, Growing Environment, and Genotype on the Contents of Selected Phenolics in Strawberry (Fragaria × ananassa Duch.) Fruits. Journal of Agricultural and Food Chemistry. 2006;54:2614-2620.10.1021/jf052947w

[94] Perkins-Veazie P, Roberts W, Collins JK. Lycopene Content Among Organically Produced Tomatoes. Journal of Vegetable Science. 2007;12:93-106.10.1300/J484v12n04_07

[95] Barrett DM, Weakley C, Diaz JV, Watnik M. Qualitative and Nutritional Differences in Processing Tomatoes Grown under Commercial Organic and Conventional Production Systems. Journal of Food Science. 2007;72:C441-C451.10.1111/j.1750-3841.2007.00500.x

[96] Wang SY, Chen C-T, Sciarappa W, Wang CY, Camp MJ. Fruit Quality, Antioxidant Capacity, and Flavonoid Content of Organically and Conventionally Grown Blueberries. Journal of Agricultural and Food Chemistry. 2008;56:5788-5794.10.1021/jf703775r

[97] Stracke BA, Rüfer CE, Weibel FP, Bub A, Watzl B. Three-Year Comparison of the Polyphenol Contents and Antioxidant Capacities in Organically and Conventionally Produced Apples (Malus domestica Bork. Cultivar `Golden Delicious'). Journal of Agricultural and Food Chemistry. 2009;57:4598-4605.10.1021/jf803961f

[98] Luthria D, Singh AP, Wilson T, Vorsa N, Banuelos GS, Vinyard BT. Influence of conventional and organic agricultural practices on the phenolic content in eggplant pulp:

Plant-to-plant variation. Food Chemistry. 2010;121:406-411.http://dx.doi.org/10.1016/j.foodchem.2009.12.055

[99] Balisteiro DM, Rombaldi CV, Genovese MI. Protein, isoflavones, trypsin inhibitory and in vitro antioxidant capacities: Comparison among conventionally and organically grown soybeans. Food Research International. 2013;51:8-14.http://dx.doi.org/10.1016/j.foodres.2012.11.015

[100] Miranda Rossetto MR, Shiga TM, Vianello F, Pereira Lima GP. Analysis of total glucosinolates and chromatographically purified benzylglucosinolate in organic and conventional vegetables. LWT - Food Science and Technology. 2013;50:247-252.http://dx.doi.org/10.1016/j.lwt.2012.05.022

[101] López A, Fenoll J, Hellín P, Flores P. Cultivation approach for comparing the nutritional quality of two pepper cultivars grown under different agricultural regimes. LWT - Food Science and Technology. 2014;58:299-305.http://dx.doi.org/10.1016/j.lwt.2014.02.048

[102] Valverde J, Reilly K, Villacreces S, Gaffney M, Grant J, Brunton N. Variation in bioactive content in broccoli (Brassica oleracea var. italica) grown under conventional and organic production systems. Journal of the Science of Food and Agriculture. 2015;95:1163-1171.10.1002/jsfa.6804

[103] Migliori C, Di Cesare LF, Lo Scalzo R, Campanelli G, Ferrari V. Effects of organic farming and genotype on alimentary and nutraceutical parameters in tomato fruits. Journal of the Science of Food and Agriculture. 2012;92:2833-2839.10.1002/jsfa.5602

[104] Del Rio D, Rodriguez-Mateos A, Spencer JPE, Tognolini M, Borges G, Crozier A. Dietary (Poly)phenolics in Human Health: Structures, Bioavailability, and Evidence of Protective Effects Against Chronic Diseases. Antioxidants & Redox Signaling. 2012;18:1818-1892.10.1089/ars.2012.4581

[105] Stan SD, Kar S, Stoner GD, Singh SV. Bioactive food components and cancer risk reduction. Journal of Cellular Biochemistry. 2008;104:339-356.10.1002/jcb.21623

[106] Vincent HK, Bourguignon CM, Taylor AG. Relationship of the dietary phytochemical index to weight gain, oxidative stress and inflammation in overweight young adults. Journal of Human Nutrition and Dietetics. 2010;23:20-29.10.1111/j.1365-277X.2009.00987.x

[107] Garaguso I, Nardini M. Polyphenols content, phenolics profile and antioxidant activity of organic red wines produced without sulfur dioxide/sulfites addition in comparison to conventional red wines. Food Chemistry. 2015;179:336-342.http://dx.doi.org/10.1016/j.foodchem.2015.01.144

[108] Unal K SD, Taher M. Polyphenol content and antioxidant capacity in organically and conventionally grown vegetables. Journal of Coastal Life Medicine. 2014;2:864-871, http://www.jclmm.com/qk/201411/6.pdf

[109] Straus S BF, Turinek M, Slatnar A, Rozman C, Bavec M. Nutritional value and economic feasibility of red beetroot (Beta vulgaris L. ssp. vulgaris Rote Kugel) from different production systems. African Journal of Agricultural Research. 2012;7:5653–5660, http://www.oxfordjournals.jurnalpedia.academicjournals.org/article/article1380984270_Straus%20et%20al.pdf

[110] Oliveira AB MC, Gomes-Filho E, Marco CA, Urban L, Miranda MRA. The Impact of Organic Farming on Quality of Tomatoes Is Associated to Increased Oxidative Stress during Fruit Development. Plos One. 2013;8.10.1371/journal.pone.0056354

[111] Lamperi L, Chiuminatto U, Cincinelli A, Galvan P, Giordani E, Lepri L, et al. Polyphenol Levels and Free Radical Scavenging Activities of Four Apple Cultivars from Integrated and Organic Farming in Different Italian Areas. Journal of Agricultural and Food Chemistry. 2008;56:6536-6546.10.1021/jf801378m

[112] Faller ALK, Fialho E. From the market to the plate: Fate of bioactive compounds during the production of feijoada meal and the impact on antioxidant capacity. Food Research International. 2012;49:508-515.http://dx.doi.org/10.1016/j.foodres.2012.08.008

[113] Aldrich HT, Salandanan K, Kendall P, Bunning M, Stonaker F, Külen O, et al. Cultivar choice provides options for local production of organic and conventionally produced tomatoes with higher quality and antioxidant content. Journal of the Science of Food and Agriculture. 2010;90:2548-2555.10.1002/jsfa.4116

[114] Mulero J, Pardo F, Zafrilla P. Antioxidant activity and phenolic composition of organic and conventional grapes and wines. Journal of Food Composition and Analysis. 2010;23:569-574.http://dx.doi.org/10.1016/j.jfca.2010.05.001

[115] Heimler D, Vignolini P, Arfaioli P, Isolani L, Romani A. Conventional, organic and biodynamic farming: differences in polyphenol content and antioxidant activity of Batavia lettuce. Journal of the Science of Food and Agriculture. 2012;92:551-556.10.1002/jsfa.4605

[116] Borguini RG, Bastos DHM, Moita-Neto JM, Capasso FS, Torres EAFdS. Antioxidant potential of tomatoes cultivated in organic and conventional systems. Brazilian Archives of Biology and Technology. 2013;56:521-529, http://www.scielo.br/scielo.php?script=sci_arttext&pid=S1516-89132013000400001&nrm=iso

[117] Navarro P, Pérez-López AJ, Mercader MT, Carbonell-Barrachina AA, Gabaldon JA. Antioxidant Activity, Color, Carotenoids Composition, Minerals, Vitamin C and Sensory Quality of Organic and Conventional Mandarin Juice, cv. Orogrande. Food Science and Technology International. 2011;17:241-248.10.1177/1082013210382334

[118] Roitner-Schobesberger B, Darnhofer I, Somsook S, Vogl CR. Consumer perceptions of organic foods in Bangkok, Thailand. Food Policy. 2008;33:112-121.http://dx.doi.org/10.1016/j.foodpol.2007.09.004

[119] Cerjak M, Mesić Ž, Kopić M, Kovačić D, Markovina J. What Motivates Consumers to Buy Organic Food: Comparison of Croatia, Bosnia Herzegovina, and Slovenia. Journal of Food Products Marketing. 2010;16:278-292.10.1080/10454446.2010.484745

[120] Ozguven N. Organic Foods Motivations Factors for Consumers. Procedia - Social and Behavioral Sciences. 2012;62:661-665.http://dx.doi.org/10.1016/j.sbspro.2012.09.110

[121] Ballute AK BP. The perceptions of and motivations for purchase of organic and local foods. Journal of Contemporary Issues in Business Research. 2014;3:1-18, http://jcibr.webs.com/Archives/Volume-2014/Issue-1-january/Article-V-3-N-1-082013JCIBR0037.pdf

[122] Henryks J PD. Investigating the context of purchase choices to further understanding of switching behaviour. Journal of Organic Systems. 2014;9:38-48, http://www.organic-systems.org/journal/92/JOS_Volume-9_Number-2_Nov-2014_Henryks-&-Pearson.pdf

[123] Stanton JV, Guion DT. Perceptions of "Organic" Food: A View Through Brand Theory. Journal of International Food & Agribusiness Marketing. 2015;27:120-141.10.1080/08974438.2014.897667

[124] Cicia G DGT, Ramunno I, Tagliaferro C. Splitting Consumer's Willingness to Pay Premium Price for Organic Products over Purchase Motivations. 98th Seminar of the European Association of Agricultural Economics (EAAE) Marketing Dynamics within the Global Trading System: New Perspectives, Chania, Crete, Greece, June 29 - July 2. 2006, http://www.researchgate.net/profile/Pietro_Pulina/publication/28685241_The_Motivational_Profile_of_Organic_Food_Consumers_a_Survey_of_SpecializedStores_Customers_in_Italy/links/00b7d52dfdeef516f5000000.pdf

[125] Monk A M, B, Lobo A, Chen J, Bez N. Australian organic market report 2012. Brisbane: Biological Farmers Association (BFA) Ltd. 2012, 100 pp. http://austorganic.com/wp-content/uploads/2013/09/Organic-market-report-2012-web.pdf

[126] McCarthy B, Murphy L. Who s buying organic food and why?: Political consumerism, demographic characteristics and motivations of consumers in North Queensland. Tourism & Management Studies. 2013;9:72-79, http://www.scielo.mec.pt/scielo.php?script=sci_arttext&pid=S2182-84582013000100011&nrm=iso

[127] H I. Consumers' Attitude and Intention towards Organic Food Purchase: An Extension of Theory of Planned Behaviour in Gender Perspective. International Journal of Management, Economics and Social Sciences. 2015;4:17 – 31, http://ssrn.com/abstract=2578399

2

Alternative Foods — New Consumer Trends

Mehdi Zahaf and Madiha Ferjani

Additional information is available at the end of the chapter

Abstract

Increased globalization of food systems, large-scale production and distribution, and retail sales have changed the way food is produced and consumed. The dis-embedded globalized system is characterized by "industrial food" and not well-informed food choices.
This has also created many concerns with respect to food safety, food security, health,
and sustainability. Food alternatives are developing leading to embedded localized systems. These "alternative food" options include labels such as local, natural, pesticide-free,
ecologically friendly, slow food movement, and localvores. The traditional marketing approach and specifically consumer marketing theory are not sufficiently prepared to han-
dle the advent of new types of consumers. These consumers are looking for more than a
product, i.e., value products. The objective of the current study is to understand the motives and concerns, product preferences, and consumption patterns of alternative food
consumers in both developed and developing countries. To this end, a survey was administered in two countries. The population targeted for this study is alternative food
shoppers. Results show mitigated differences between developed country consumers and
developing country consumers in terms of food culture and food importance, perception
of organic versus local foods, and foods channels of distribution.

Keywords: Organic food, local food, consumer behaviour, distribution

1. Introduction

1.1. New food market realities

The last two decades were driven by two major trends in the agriculture industry: an increase
in the use of genetically modified food (GMF) and an increase in food-related diseases, such
as mad cow, bird flu, and more recently the horsegate [33]. Emerging efforts to provide food
safety and quality has led to a grown number of quality assurance schemes both at national
and international levels. To this end, several "new" alternatives eliminate a number of concerns

towards industrial food production and distribution. These "alternative foods" options include labels such as local, natural, organic, and more recently, paleo. Advocates of these movements are against any industrialization of the food chain, its production, and distribution. This system is based on two major elements, namely: (i) food mileage and carbon footprint and (ii) non-industrialization of the food chain. It is obvious that support for the local economy and country of origin are by-products of such system.

The organic market moved from a niche market to a mainstream market in the last two decades. This trend originated in the nineties, following a number of food scares in the conventional sector. The global market for organic products was approximated at US $18 billion in 2000, then US $23 billion in 2002, then increased by 43% reaching US $33 billion in 2005, and US $50 billion in 2008 [40, 36]. In the last decade, double-digit growth rates were observed each year [41]. Further, there are 633,891 farms managing 31 million hectares of "organic" land [40]. Although organic agriculture is now going mainstream, its credibility might be jeopardized as the production methods and processes are being industrialized [4]. Padel and Foster [26] claim that *"Although demand for organic food is still buoyant, there are signs that markets are maturing and growth rates over the last years have slowed to below 10%"*. The main critics are not related to the key elements in the current definition of organics. On the contrary, these concerns are directly related to some economic, environmental, and social ideals such as production systems, size of the operations, distribution systems and channels, and capital intensity. The by-product of this situation is what Bean and Sharp [4] call alternative food systems (AFS). These systems are sustainable and economically, socially, and environmentally more viable. Concepts such as local, fair trade, and paleo come into play here.

2. Alternative foods

2.1. Variety and food labels

Aside from hardcore consumers that are very knowledgeable, others are still not well educated about the meaning of alternative food labels. Although there is a lack of a widely accepted single definition of these new alternative food concepts, there are serious attempts to provide clear bounds to this label. In fact, radial distance, such as 100 miles, replaced ambiguous characteristics such as political lines of distinction [39] or distinct characteristics of people and places [3]. In addition, Geographical Indication Labels (GIs) provide a clear signal to identify a local product. The European Union, for instance, recognizes two basic categories of GIs: the Protected Designation of Origin (PDO) and the Protected Geographical Indications (PGI). These labels help consumers not only recognize where the product comes from but also the production methods used [15].

The use of the term "organic" is restricted to farms, products, processors, and other intermediaries in the value chain between production and consumption, which have been certified by Certifying Bodies. The USDA[1] provides organic labeling to *"products raised without the use of most conventional pesticides, petroleum or sewage-based fertilizers, or genetically engineered materials"*, in addition to the use of renewable resources and conservation. "Transitional organic" is also a restricted label and describes farms which have made the commitment to move toward organic certification. According to an FiBL[2] survey on organic rules and regulations, there are 82 countries with organic regulation and 16 countries in the process of drafting legislation [10]. In the same report, the organic sector is considered as the linchpin to face the challenges of food security, climate change, poverty alleviation, hunger, health, and biodiversity steward-ship. Since the principles of organic agriculture include issues of social justice, Browne et al. [7] noted that sustainability and organics are closely linked and that ethical and organic trading are beginning to overlap.

Besides ensuring no use of genetic engineering, pesticides, additives, or fertilizers, local food labels should provide the consumer the value related to operation size, as well as distribution. In other words, buying local food should contribute to protecting the local farming economy, as well as the environment by reducing "food miles". In addition, culture is another important dimension which might be considered in defining local foods. Besides associating terroir and local food products with PGI, PDO, TSG (Traditional Specialty Guaranteed), food baskets, distributor's own label, or slow food, Bérard and Marchenay [5] underline the concept of localized food, which is based on the cultural dimension [20]. Consumers, particularly locavores, are becoming considerate not only about where their food comes from and pro-duction processes but also the way the food is made and creative versions of regional food classics of each season [12]. That said, it is important to consider what consumers qualify as "locally grown" since it determines differentiation patterns and, consequently, profits [9].

Labels like "local", "natural", "paleo", "pesticide-free", and "ecologically friendly" are not regulated and tend to be used by small farms catering to local or regional clientele. With the exception of marketing board-regulated products like dairy or chicken, production and handling of foods sold under these labels are for the most part not monitored or regulated, except by governmental agencies and district health units, and then only in terms of health/ safety inspections and only as required by law. As a result, information on farms operating outside of the organic certification system is scattered and incomplete. Lastly, "organic" foods have to be differentiated from "functional" foods [35]. Organic foods tend to be regulated and are based on supply side value while functional foods are not very regulated and are based on demand side value. While both types of product are marketed to achieve the same objective (i.e., healthy products), the market positioning is very different.

2.2. Motivations and reasons to buy

Studying what determines consumer preferences for local food, as well as organic food, has been the concern of numerous studies in different countries [26, 6].

Aprile et al. [2] piloted a segmentation analysis of olive oil consumers in order to analyze consumers' attitude towards local produce in Naples, Italy. They identified four clusters of local food consumers: local traditionalist, local ecologist, local fans, and local health conscious. Results show that seven factors explain consumer attitudes towards local food consumption:

---

1 United Stated Department of Agriculture.

2 The Research Institute of Organic Agriculture (FiBL)

health concerns, altruism, environmental concerns, local habitual, local origin, certification, and specialties. Willingness-to-pay for PDO and PGI labels and other quality signals vary across the different identified segments. Similarly, Aguirre [1] conducted a comparative synthesis of the organic consumer profile in four different locations, US, Canada, Europe, and Costa Rica, based on three criteria: socio-demographics, purchase motivations, and main concern. The results indicate important similarities among the US, Canada, and Europe organic consumer with the Costa Rican consumer. Particularly in the four locations, the purchase motivations relate to health, environment, no-use of chemical, some concern about ethical issues, and helping farmers. Despite some differences in the barriers to purchase, consumers in all four locations state factors such as price and availability or unstable supply.

The importance of consuming local food is increasingly converging across different countries and cultures. Green et al. [16] conducted a study in four European countries (Finland, Germany, Italy, and UK) and the results of the study reveal the relative importance of risk associated with consuming conventional industrialized food, as well as the issue of provenance of food as a key element of the cultural framework in all countries. This highlights the fact that consumers seek alternative food as a way to reduce this risk and the importance of trust to facilitate choices in complex choice situations. Consequently, in making complex decision choices, consumers tend to use "pragmatic decision aids" rooted in cultural frameworks, as well as "craft skills", in order to assess food quality [16].

When it comes to understanding the main reasons for organic food consumption, Tarkiainen and Sundqvist [34] suggest that it is a way of life connected to a particular value system that affects attitudes, and consumption behavior. Padel and Foster [26] tried to ascertain those underlying values taking into account differences among consumers in terms of frequency of purchase and demographics (gender, marital status, number of children, etc.). Those values include enjoyment, unity with nature, respect for nature, taking care of family, benevolence, etc. More specifically, organic food-sales volume increase is due to consumers' self-interest motives that are predominant (e.g., personal health, high food quality, and taste). These are widely cited in the literature as the key factors to explain consumers' purchasing decision of organic food [24, 42]. However, it has been argued that organic food consumers might also have altruistic motives (e.g., environmentally friendly, animal welfare, fair trade). In Canada, organic food consumers mainly identify health and the environment, as well as support of local farmers, as main motives for their food consumption [19]. In the same vein, the Norm Activation Theory [29] explains altruistic behavior by feelings of moral obligation to act on one's personal internalized norms. This theory is particularly relevant in explaining consumers' attitudes towards organic food as an ethical food choice, which is based on political, ecological, and religious motives [21]. These political motives confirm Weber's [37] statement that human behavior is a way to affirm oneself and differentiate social status and belonging to groups.

Overall, growing consumer demand for alternative foods has been attributed to consumers' concerns regarding nutrition, health, the environment, and the quality of their food [14, 23, 31]. Further, various studies conducted in Europe and the US have explored consumer behavior and have tackled the issue of determining consumers' motivations and preferences for organic products [42, 38]. Although some consumers are environmentally conscious, most studies confirm the predominance of egocentric values like health, attitude towards taste, and freshness that influence alternative food choices [13, 42]. That said, Padel and Foster [26] show that motives and barriers may change with the purchasing frequency and across product categories. They distinguish between regular consumers who are generally families with at least one child suffering from asthma or food allergies and non-buyers who are more skeptical about organic food benefits and more sensitive to price premiums. They also highlight that consumers consider fruits and vegetables as the "key entry points" to the "organic experience", followed by other categories such as eggs and dairy, grocery products, meats, and soft drinks. In addition, their study reveals that trust appears as an important factor in deciding where to buy. In fact, consumers trust more specialist organic or local shops rather than supermarkets and large corporations.

On the other hand, the main reasons that prevent consumers from buying alternative foods are expensiveness, limited availability, unsatisfactory quality, lack of trust, lack of perceived value, poor presentation (packaging, display) and misunderstanding of the production processes, and lack of information [13, 14, 23]. In fact, the lack of information is related to the ability of consumers to locate organic products, to learn about the organic certification process, in addition to their ability to identify an organic product. The easiest way is to look for the word "organic" on the label. However, some consumers are familiar with various organic labels and might choose based on other features such as "natural". Conversely, previous research on the recent growth of consumer interest in local food shows that it is attributed to increased concerns with safety and accountability about food, in addition to a desire to support regional farmers, the local economic and natural environment. Consumers want to know where their food comes from and how it is grown or raised.

2.3. Global versus local production and distribution

With the rapid growth of the organic supply, producers moved from traditional production methods to more industrialized production methods. Industrial farming addresses efficiently and effectively the challenges related to the cost and logistics of moving produced foods to national and global markets. Conventional food value chain applies an important downward pressure on price leading to the issues of profitability and productivity. This has resulted for some small farmers - concerned with the philosophical aspects of organic production – indiminished credibility of the organic standard and in a refusal to industrialize. These key contradictions lead to a "bifurcation" between market- and movement-oriented organic distribution systems since dedicated consumers continue to support alternative organic networks [28]. It has also hardened the value chain against entry by these small farmers. Hence, the challenge that the alternative food system is facing is a gap that spans between the consumerism/producerism system in place, the current food chain, and the alternative value delivery network/value chain.

Furthermore, this gap is broader between developed and developing countries. It is interesting to shed the light on similarities and differences between developed and developing countries in terms of the variables that might shape the buying behavior of organic foods consumers versus local foods consumers. As a matter of fact, there were almost 1.9 million organic producers in 2009, an increase of 31% since 2008, mainly due to a large increase in the production in India. Further, 40% of the world's organic producers are in Asia, followed by Africa (28%), and Latin America (16%). In North America, Canada allocates 0.7 million hectares to organic production while the United States has 2 million hectares. This represents 7% of the world's organic agricultural land.

One could infer that developing countries are increasingly concerned about providing food safety and all the ecological, social, and economic motivations behind adopting this option. However, some studies proved that *"the main aim of several developing countries' policies and/or legislative approaches for organic agriculture is income generation through the promotion of certified organic food"* [30]. In Tunisia, for instance, the Tunisian government developed policies, established a National Commission for Organic Agriculture and a certification authority, assigned a budget to cover 30% of investments of organic farmers and 70% of certification costs over five years to encourage farmers' conversion to organic production to comply with EU Regulation since 1999. Those incentives made Tunisia ranked 35th worldwide, and the 1st among African countries, in terms of certified area (87,000 hectares). An interesting aspect to grasp is the role of these institutions in promoting and educating Tunisian consumers about organic food.

3. Conceptual framework

The approach of the current study is based on an integrative production-distribution-consumption model (cf. Figure 1). There are three layers of decision in this model: (i) supply chain related to certification and production methods; (ii) value delivery network related to the channels of distribution broken down into three main categories, long or standard channel, short channels, and direct channels; and finally, (iii) the consumer behavior related to the psychographics influencing the consumption of alternative food.

The tri-Party model shows the alternative food value that will be assessed in this study. Basically, consumers are assumed to have a certain *food culture* that is directly related to the degree of economic development. This in turn sets the current standard of food production that leads ultimately to food concerns. These concerns will—again—influence the way consumers perceive and eat food (food culture). Consequently, these perceptions give rise to food preferences and, more importantly, reasons to buy and requests regarding food quality, freshness, environmental and economic impacts, and healthiness. This is assumed to depict a certain size of operations (large versus small). This in turn will impact the type of channel members involved in these operations. It is assumed here that the distribution channels are very short, counting a maximum of two members: one producer/farmer and one distributor (if there are any). These channels create values that are logically different depending on the point of sale. Lastly, depending on the market coverage and the channel size, farmers, producers, or distributors will have a marketing approach adapted to the value offered to the target market.

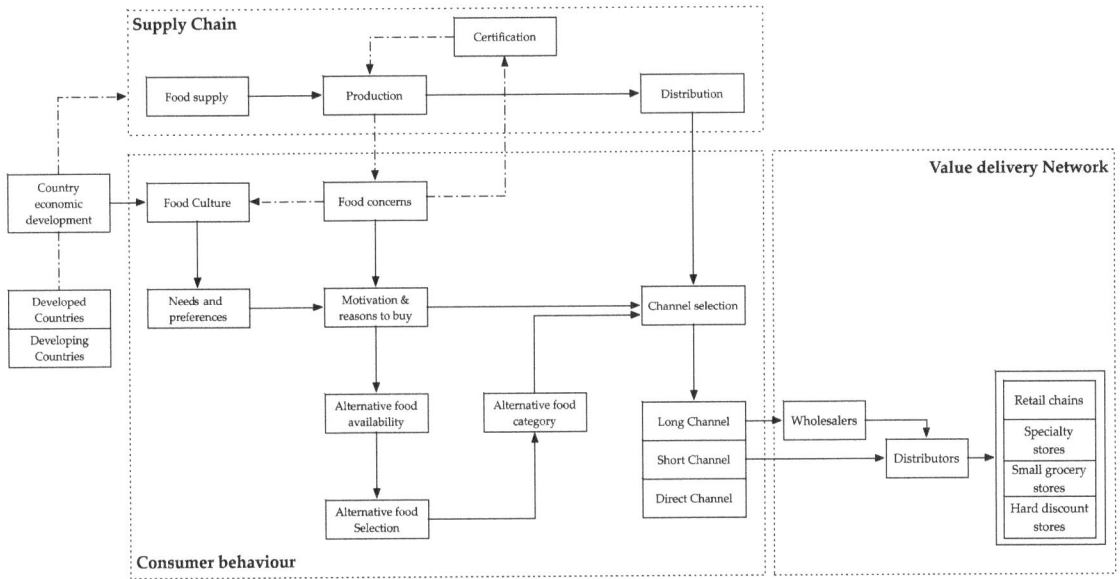

Figure 1. Integrative Production-Distribution-Consumption Model

4. Research design

4.1. Objectives

The current study aims to uncover the demand and supply side factors that affect the alternative foods supply chain and how value is created through the distribution channel and perceived by the final consumers. This value needs to be determined and estimated at the demand side level. Further, the logistics of the value delivery network need to be investigated. This will lead to an in-depth understanding of the value added in the alternative food distribution system, the current market structure, as well as its determinants. Further, building trust in the organic food (OF) supply requires more than just ensuring product quality and product knowledge, or labeling and setting proper pricing and communication strategies, as actually trust is missing at various levels of the marketing value delivery system and the food supply chain. The dimensions of trust necessary to achieve market growth have to be integrated to the OF product positioning and the distribution strategies. Moreover, this will provide a detailed assessment of the actual purchasing situation in the current distribution system, e.g., superstores, specialty stores, and farmers' market. This analysis is done taking the perspective of both a developed country (Canada) and a developing country (Tunisia). This will help to understand the importance of the value delivery network in creating value added to the target market. Hence, the second objective is to explore the market responsiveness to the different distribution strategies used in developed and developing countries. In order to target more efficiently consumers, we need to provide a more precise and useful profile of these consumers, who they are, what they eat, how they buy, where they buy, and why they eat alternative foods. This will lead to an in-depth understanding of the major forces shaping the current market structure, as well as an understanding of the challenges faced by the main players of the alternative food industry.

Hence, our objectives can be summarized as follows:

1. Determine alternative food consumers' purchasing behavior in terms of how consumers buy, where they buy, reasons to buy, attitudes, expertise, and trusted channels of distribution;

2. Compare consumers' purchasing patterns of developed and developing countries; and

3. Cluster alternative food consumers with regard to their psychographics in both country types.

4.2. Operational framework

This operational model shows the alternative foods value that will be assessed in this study. Basically, as it is shown in Figure 2, consumers are assumed to have requests and preferences regarding food quality, freshness, environmental and economic impacts, and healthiness. This is assumed to depict a certain size of operations (country economic development). This in turn will impact the expertise and familiarity of these consumers with regard to alternative foods. These elements are the foundation of the motivation to buy alternative foods.

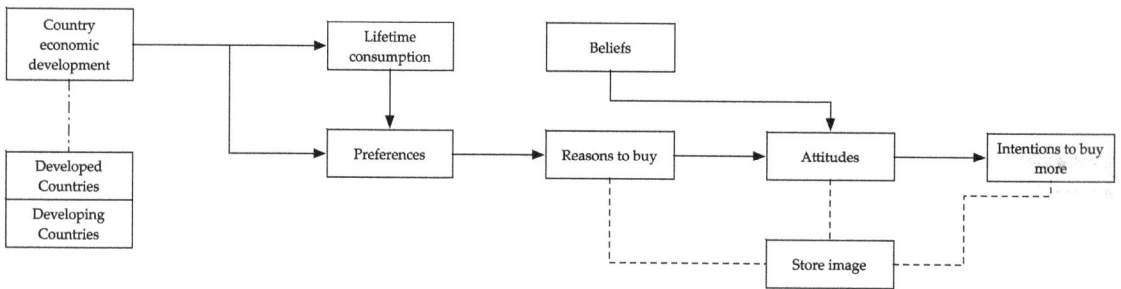

Figure 2. Operational Framework

Preferences will drive the motivation to buy alternative foods. It is assumed that beliefs, motivation, and attitudes are prerequisites to intentions to buy. Lastly, store image as defined above plays a moderating role here.

4.3. Measurement and scaling

To address the study objectives, a quantitative design is required. The design will help profile consumers by country and their purchasing patterns. The conceptual framework depicted in Figure 1 has been developed to assess the alternative food consumption schemes. This in turn is expected to lead to the development of a second model that also takes into account the key factors shaping this new market. The former model has been tested using a structured questionnaire. Prior to developing the survey, secondary data was collected in Canada and Tunisia using major sources of information, as well as informal interviews with industry key players (experts, certifiers, and government representatives). These gatekeepers can provide the most recent and accurate information about the alternative food market and industry.

Information obtained from these key players, while fairly comprehensive within its scope, is not necessarily accurate. This is illustrated by the example that in order to reach various target export markets, some farms, products, and businesses are certified by multiple bodies simultaneously.

The output of these interviews helped design the questionnaire. This latter is structured into three sections. The first section deals with consumers' general opinion about organic food, consumption and shopping habits, and reasons for buying organic products (measured on a 5-point Likert scale). The second section of the survey measures consumers' psychographics in terms of trust, beliefs, and attitudes (all measured on a 5-point Likert scale). Finally, the third section is structured to design a socio-demographic profile of our respondents. The survey was developed by selecting other case study questionnaires on the topic of alternative food marketing [27, 11, 32, 13, 17]. Prior to administering the survey, a pre-test was done and minor modifications were made. Quantitative data for this study has been analyzed using the Statistical Package for the Social Sciences (SPSS). A total of 500 questionnaires were collected, and 480 questionnaires were usable. Data was cleaned and missing values were replaced using the mean. All variables were tested to check their internal consistency. Further, all reliability tests were coupled to a series of factor analyses to determine the structure of the data. Factor analyses also helped to test if the items were measuring the right constructs. Results from factor analysis and reliability analysis show good levels for an exploratory study [18].

4.4. Sampling design

To address the abovementioned objectives, alternative food consumers have been surveyed to assess their consumption behavior/patterns. Hence, a survey was administered to consumers in a developed country (Canada) and a developing country (Tunisia). The population targeted for this study is alternative food shoppers (organic food, certified organic food, local food, and fair trade food). For the purpose of gaining a good representation, respondents needed to fit within a specific profile. The idea was to randomly select alternative food consumers that make their purchase mainly at small producers' farm gates, community farmers, farmers' market, community groceries, specialty stores, and community chain stores. Further, they had to consume at least one of the following product categories: fruits, vegetables, dairy, bread, meat, and prepared food. They also had to be in charge of household grocery/food purchases. This being said, countries have been selected based on the stage of alternative food product's life cycle. Further to this, it is well known that food is culture in developing countries while in developed countries, this is not the case [5].

The point of contact of data collection—point of respondent interception—was selected according to the value delivery network. It is obvious that developing countries present different marketing distribution patterns than developed countries. More precisely, the delivery chain differs as per (i) channel size and type, (ii) alternative food products variety, and (iii) channel position—number of layers in the distribution system. Developed countries align all types of channels of distribution while developing countries have limited distribution channels embodied mainly in the direct channels (producers) and, to a limited extent, in short channels (specialty stores). Lastly, there is a two-prong challenge related to surveying some of these distribution players: (i) limited availability of some alternative food, and (ii) the limited size of the population requires a large sample size sufficient enough to ensure consistency of the results without reaching any saturation.

5. Results

5.1. Overall consumers profile

Consumers have been profiled using the data collected from the respondents who indicated that they currently purchase alternative foods (mainly organic and local). Overall, the typical alterative food consumers are aged 25 to 35 years old (30.1%); single (63.3%); household composed of 4 to 5 persons (38.6%); have at least an undergraduate degree (51.5%); buy at least two organic food products (90.8%); eat mainly national country-based organic (32.1%); buy organic food mainly from supermarkets; and finally, consider price as the major determinant when buying alternative foods.

5.2. Lifetime consumption: Familiarity and expertise

Consumers have been regrouped using their lifetime consumption. As per Cunningham's [8] work, if respondents have been buying alternative foods on a regular basis, then they are classified as regular alternative food consumers (RAFC); while if they haven't been consuming alternative foods for a very short period of time, then they are tagged as non-regular alternative food consumers (non-RAFC). It is important to note here that alternative foods have been defined in broad terms of consuming either organic foods (certified, fair trade, local) or local (foods). Accordingly, respondents are distributed as follows: 63.1% of RAFC and 36.1% of non-RAFC. This means that a third of the consumers has been consuming alternative foods for more than a year while the rest of the sample have shorter experience with the product. Lastly, RAFC and non-RAFC are almost equally distributed on the Canadian sample, while in the Tunisian sample there are more non-RAFC (76.9%) than RAFC (23.1%).

Lifetime consumption could serve as a proxy to several indicators such as experience with the product, knowledge about the points of sales and price differentials, and level of trust. To corroborate this, several ANOVAs were run to check if there are significant differences between RAFC and non-RAFC in terms of their familiarity and expertise with regard to alternative foods. Results show that RAFC are more familiar than expert when compared to non-RAFC. These findings are summarized in Table 1.

	RAFC	Non-RAFC	Significance level
Familiarity with alternative foods	4.68	3.21	0.000*
Expertise	3.76	2.40	0.000*

Table 1. Familiarity and Expertise of RAFC and non-RAFC

5.3. Purchasing pattern

5.3.1. Purchase criteria and preferences

Given that the survey did not clearly define what alternative food is, it is assumed that respondents understand this concept. Further, there was no differentiation between local, local organic, fair trade organic, and certified organic. This is also evidenced by how respondents addressed the question related to alternative food preferences. In terms of local food consumption, 21.9% of respondents indicated they do purchase local organic food, 32.1% purchase national organic (Nationally produced – Canada or Tunisia), 7.3% buy certified organic, 4.3% buy fair trade organic foods, and 33.8% have no specific preference.

Attributes	Canada	Tunisia
National organic	11.9%	20.9%
Certified organic	2.3%	5%
Local organic	15.4%	6.5%
Fair trade organic	0.8%	3.8%
No preference	21.3%	12.3%

Table 2. Cross Tabulation: Country versus Product Preferences

Table 2 shows that RAFC and non-RAFC are mainly looking for the national and/or local food dimension. This downgrades certification and fair trade to lesser importance. These consumers are more hardcore alternative food consumers looking for good value products.

Further, when classifying these results by country, it is clear that consumers in developing countries do not clearly differentiate between the different types of alternative foods. This is mainly due to cultural food factors; the agricultural sector is not industrialized yet in developing countries. Consumers tend to associate agricultural production to local/national production. Imports are not as important as for developed countries. This is evidenced by the Chi-square test. It shows that there is an association between the country and alternative food preferences (χ^2=53.88, p=0.000).

Furthermore, a simple mean analysis[3] shows that the three most important criteria when buying alternative foods are: healthiness (4.79), quality (4.79), and support to the local economy (4.81). Taste and environmental friendliness do not seem to be important purchasing criteria (mean lower than 1). Moreover, RAFC show higher means on the five dimensions than the non-RAFC. However, the only significant differences are related to taste and environmental friendliness. This shows again that regardless of their familiarity and expertise, the most important factors for consumers are intrinsic attributes (healthiness and quality) and extrinsic attributes (support to the local economy).

---

3 On a five-point Likert scale.

5.3.2. Point of purchase

Question 10 of the survey measures consumers' perception of the store offering and value. This is a very important indicator of the store impact on consumers' choices. Table 3 shows that all dimensions are relatively important to all consumers; quality, convenience and services being the most important factors. Price is moderately important and presents the lowest score (3.51). The mode for all dimensions is 4 on a scale of 5. Hence, all criteria are considered by consumers but to different extents when buying alternative foods.

	Mean	Mode
It is convenient to do my shopping in this store	3.68	4
It offers a wide variety of products	3.60	4
It offers good quality products	3.82	4
It offers the services I am looking for	3.67	4
It offers good prices	3.51	4

Table 3. Store Choice Mean Analysis

To complement these analyses, bivariate correlations were run to show that store choice is related to intentions to buy, attitudes, and reasons to buy. This proves the homogeneity and structure of the purchase behavior.

Lastly, an ANOVA was run to check if there are differences between developed and developing countries in terms of store choice. Results are not conclusive. However, even though there is no significant difference between both countries, it is interesting to note that consumers in developed countries have higher scores on all dimensions than developing countries. This clearly shows that the former countries have a stronger store image than the latter countries. This is mainly related to the degree of economic development and the structure and maturity of the value delivery network.

5.3.3. Buying process

In the current study, the buying process is measured with a multi-step sequence starting with motivations, beliefs, reasons to buy, and ending with intentions to buy more alternative foods. This latter variable is dependent on attitudes that is, in turn, dependent on beliefs and reasons to buy. Attitudes are considered as a proxy for the final purchasing behavior. Two simple linear regressions were run to test the buying process. Before running the first regression, a factor analysis was run to determine the number of dimensions of the variable beliefs towards alternative foods. Results show two dimensions: one related to the intrinsic attributes such as taste and healthiness, and another one related to the extrinsic attributes such as price and the meaning of alternative foods.

Regression 1 tests the influence of the reasons to buy and beliefs (intrinsic and extrinsic) on attitudes (cf. Table 4).

Independent	Sig.	Beta
Reasons to buy	0.000*	+0.394
Intrinsic beliefs	0.000*	+0.305
Extrinsic beliefs	0.069	-0.049

Table 4. Regression 1: Reasons and Beliefs on Attitudes

Reasons to buy and intrinsic beliefs are determinants of attitudes. Both explain 33.5% of the variance of this latter variable and both have a positive influence on attitude. It is important to note that consumers do not consider extrinsic beliefs when building their attitudes. This shows clearly that such consumers look more for a value rather than a product. Regression 2 tests the last link in the process, namely the influence of attitudes on the intentions to buy more alternative food products (cf. Table 5).

Independent	Sig.	Beta
Attitude	0.000*	0.699

Table 5. Regression 2: Reasons and Believes on Attitudes

As expected, attitudes have a positive effect on intentions to buy more alternative foods (R^2=32.1%). To recapitulate, Regressions 1 and 2 show that there is a linear relationship between reasons to buy, beliefs, attitudes, and intentions to buy more alternative foods.

It is important to test whether these results hold true for both countries. Several ANOVAs have been run to test differences and similarities between Canada (developed country) and Tunisia (developing country). All results are depicted in Table 6. It is obvious that there is no significant difference between both countries in terms of reasons to buy, attitudes, and intentions to buy. However, there is a difference in terms of intrinsic and extrinsic beliefs. It is also important to note that Canadians score higher than Tunisians on all variables except for extrinsic beliefs. This is in line with the previous regression results.

Variable	Mean Tunisia	Mean Canada	Sig.
Belief – Intrinsic	3.70	3.89	**0.000***
Belief – Extrinsic	3.61	3.36	**0.007***
Reasons to buy	3.87	3.90	0.661
Attitudes	3.97	4.00	0.285
Intentions to buy more	3.72	3.77	0.543

Table 6. ANOVA Inter-country Tests

Further, all consumers score relatively higher on attitudes and reasons to buy. As expected, the lowest scores are for extrinsic beliefs. As stated in the literature review, extrinsic beliefs do make more sense for developed countries than developing countries.

5.4. Clustering consumers

Since the main focus is to classify consumers with regard to their motivation, attitudes, beliefs, expertise, and their intentions to buy more alternative foods, various analyses were run. Therefore, cluster analysis and discriminant analysis are natural techniques to segment the alternative food market and discriminate between consumers. This approach is best suited to identify consumption and behavior patterns and create a consumer typology. Specifically, we are more interested in exploring differences in behavior between the segments than predetermining the number of segments.

Different combinations of socio-demographic indicators and psychographic variables have been implemented to determine with minimal bias an optimal segmentation strategy. The idea is to maximize intra-group homogeneity and intra-group heterogeneity. This allows for more robust profiling, as consumers behave in the same way when they belong to the same segment and behave differently if they belong to different segments. Note that homogeneity and heterogeneity are defined with regard to the segmenting variables. A good segmentation is defined as a segmentation strategy that maximizes both the inter-group homogeneity and intra-group heterogeneity. Conversely, a broad segmentation is defined as a segmentation strategy that minimizes both the inter-group homogeneity and intra-group heterogeneity.

Different combinations of socio-demographic indicators and psychographic variables have been used to segment the market. Several of these combinations show problems with either the intra-group homogeneity or the inter-group heterogeneity. Alternatively, for the purpose of having a good measure of intra-group heterogeneity, several ANOVAs were run to make sure that consumers in different segments have different profiles. All tests were conclusive.

5.5. Intentions to buy more alternative foods

Our aim here is to classify respondents based on their intentions to buy more alternative foods. Question 8 prompts respondents to rate their willingness to buy more alternative foods in the future. This has been done using a five-point itemized scale, with a median point of 3. A two-step cluster analysis was run. Results show that we have a good segmentation strategy with three distinct segments (cf. Table 7).

Segments	Percentage	Mean
High intentions to rebuy	27.8%	4.86
Moderate intentions to rebuy	58.7%	3.63
Low intentions to rebuy	13.6%	2.01

Table 7. Cluster Analysis for Intentions to Buy More

Half of the consumers have moderate intention to rebuy alternative food in the future while a third of the respondents are more than willing to rebuy alternative foods in the future. Further, cross tabulations between the cluster membership and the type of alternative food consumers (RAFC–non-RAFC) show that there is an association between the type of consumers and their intentions to rebuy alternative foods. As expected, most of the high intentions to rebuy consumers are RAFC while most of the low intentions to rebuy consumers are non-RAFC.

5.6. Reasons to buy alternative foods

The two-step cluster analysis shows one cluster with high scores on the five dimensions of reasons to buy, namely healthiness, taste, environmental friendliness, quality, and support for the local economy. Factor analysis confirms one dimension for reasons to buy. A simple mean analysis[4] was run and results corroborate this finding (cf. Table 8).

	Mean	Mode
Healthiness	4.01	4
Taste	3.59	3
Environmental Friendliness	4.02	4
Quality	3.79	4
Support for the Local Economy	3.91	5

Table 8. Mean Analysis of Reasons to Buy

To investigate this finding more, several statistical checks were performed. One last cluster analysis was run to explore the effect of the country on the reasons to buy. It is interesting to see that there are two clusters intimately related to the country classification (cf. Table 9). These clusters are composed of consumers that have moderate to high reasons to buy.

	Cluster 1	Cluster 2
Cluster size	52%	48%
Clustering variable: Country	100% Canada	100% Tunisia
Clustering variable: Reasons to buy	3.90	3.87

Table 9. Cluster Analysis for Country and Reasons to Buy

5.7. Beliefs toward alterative foods

It is clear from Table 10 that true believers have positive extrinsic and intrinsic beliefs; while skeptics have the opposite beliefs. The third segment is a hybrid segment that has high intrinsic beliefs and low extrinsic beliefs.

---

4 Measured on a five-point Likert scale.

	Intrinsic Attributes	Extrinsic Attributes	Size of the Cluster
Segment 1: Skeptics	Low	Medium	29.6%
Segment 2: True believers	High	High	45.7%
Segment 3: Hybrids	High	Medium	24.7%

Table 10. Cluster Analysis for Beliefs Toward Alternative Foods

To investigate these findings more and to get plausible explanations, cross-tabulations with the type of consumers have been run (cf. Table 11). A third of the respondents are true believers and new RAFC (non-RAFC) 14.4% are RAFC. Further, there are almost three times more non-RAFC skeptics than RAFC skeptics. Lastly, there is an even distribution of non-RAFC hybrids and RAFC hybrids. These findings are in line with the results presented above. There is a strong association between the segments and the type of consumers ($x^2=14.97$, $p=0.000*$).

	Non-RAFC	RAFC
Skeptics	17.6%	6.9%
True believers	31.2%	14.4%
Hybrids	15.3%	14.6%

Table 11. Cross-tabulations of Type of Consumers and Belief Clusters

5.8. Combined clusters: Country-based clustering

Combining country and familiarity to beliefs leads to the following segments (cf. Table 12):

Segments	Acronym	Familiarity	Intrinsic beliefs	Extrinsic beliefs
Cluster 1	Tunisia	Medium	Medium	Medium
Cluster 2	Canada 1	Low	Medium	Medium
Cluster 3	Canada 2	High	High	Medium

Table 12. Cluster Analysis for a Combination of Variables

This clustering strategy shows that extrinsic beliefs are not important regardless of the country. Further, results show that there is only one cluster in Tunisia that scores medium on all variables. This could be explained by the fact that the food culture is not based on food concerns. As mentioned above, there is not industrialization of the agricultural sector. Conversely, Canada presents two opposite profiles: (i) consumers familiar with alternative food products and have expertise to assess these products,-these consumers have moderate to high beliefs; and (ii) consumers that have limited expertise regarding alternative foods, and have negative beliefs.

6. Discussion

This exploratory study has academic and practical implications to both producers/distributors and consumers. Even though alternative food has not been clearly defined in the study, results show that consumers buying local foods and fair trade or local/national organic have a purchasing behavior slightly different from what is known in the current literature. Using familiarity and expertise (lifetime consumption) as a segmentation variable provides several insights on the current behavior of RAFC. Results show that RAFC are hard-core consumers. As a matter of fact, lifetime consumption has been used as a proxy of several other psycho-graphic indicator such as trust, reasons to buy, beliefs, and intentions to buy more. Further, this adds to the classical segmentation strategy that has been used so far in the literature. For instance, compared with [22], our clustering strategy provides more insight into the why, who, and what alternative consumers buy.

Each segment exhibits a separate and distinct behavior from the other segments. RAFC are habitual purchasing consumers and non-RAFC are variety-seeking consumers. First, when buying alternative food products, RAFC are making straight habitual purchases and have their own purchasing scheme. They are characterized as consumers who are motivated by intrinsic and extrinsic attributes but only by intrinsic beliefs. This explains why these consumers have strong principle-oriented lifestyles as they also look for locally produced products and/or purchases that might help the local economy. They also care about the product quality and the healthiness. As expected, these consumers are 18 to 35, single, and educated. Gender is not determinant here; males and females exhibit the same behavior. Further to that, they buy all types of OF products ranging from fruits, vegetables, and dairy to meat. Second, non-RAFC buy alternative foods occasionally; for less than a year. For these consumers, the main reason to buy alternative foods is healthiness. However, there is a significant difference between RAFC and non-RAFC in terms of taste and environmental healthiness of alternative foods. These consumers do not perceive significant differences between alternative food and conventional food. Non-RAFC seem to have a basic trust structure. This is in accordance with [32, 19]. For instance, non-RAFC base their trust on the information available at the point of purchase because they do not collect information to build their knowledge based on OF. These consumers are not fully principle oriented.

One of the main forces that affect the current state of the market is food culture. As per Figure 1, food culture is dependent on the economic development of the country. In the context of the current study, food culture is a by-product of the industrialization or non-industrialization of the agricultural sector. In developing countries, the agricultural sector is using basic produc-tion techniques leading to the production of small quantities. These findings need to be related to the product life cycle. For instance, the organic market is driven by conventional marketing strategies and is consistently looking to standardization of the supply. This defeats the intrinsic sustainability objective of such products. This study shows the importance of the production operations and the distribution logistics. There is a clear differentiation between developed countries (using all possible distribution channels) and developing countries (using less complex distribution schemes and shorter channels). The channels reflect a certain market reality. Consumers buy from long channels because of convenience and price. They offer a local value targeted toward a certain consumer profile; these are customers that buy alternative foods for health reasons. Conversely, short channels are production method driven. These channels serve consumers that have a principle-oriented lifestyle; thus, the support of the local economy is the main drive of this market demand. Price is not an issue here.

One of the limitations of the study has been that consumers might not *fully* understand what alternative food means. Further, the analyses performed in the current study did not focus—on purpose—on the type of alternative foods. Rather, it focused mainly on (i) difference between the expertise of the consumers and (ii) differences between developed and developing countries. It would have also been interesting to study the importance of the frequency of purchase as well as price premiums. Further, the typical alternative food consumer in Canada and Tunisia is not consistent with previous research that indicates a female with a higher-level education. Having profiled this consumer, however, it is noted that consumers in both countries are very similar in terms of demographics. It is important to recognize that consumers may not fully understand the meaning of alternative foods, and thus demographics alone are not sufficient to explain the purchase behavior. Future research should be undertaken to assess the effects of different marketing ideas and also to examine if consumers understand the meaning of locally produced food.

To recapitulate, the starting point of the marketing model depicted in Figure 3 starts with the market needs. Depending on the degree of consistency of the need and the knowledge level of the target market, there are two schemes: habitual consumers (RAFC) and variety seeking consumers (non-RAFC). The more the consumers know about their needs, the more they will look for an enhanced value capturing mainly intrinsic beliefs. These consumers will look for basic channels offering quality, convenience, and services. Conversely, if consumers have limited knowledge but are driven by social consciousness (sustainability and helping the local economy), then they will buy from longer channels (specialized, community grocery stores) under the impression that food is local.

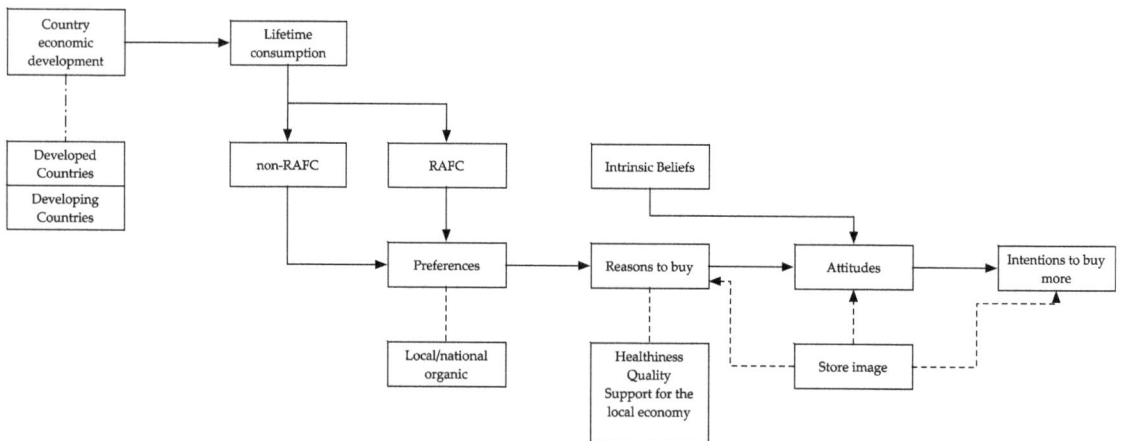

Figure 3. Final Model

7. Conclusion

Alternative food research is an area of study with a vast number of possible areas of future research. Local farmers will find value in knowing that market potential does exist for their product, and consumers are expressing an interest in purchasing locally produced food in short channels of distribution. Their motivation to buy local food products is not driven by fear and concerns over food products but rather by quality, healthiness, and support for the local economy. In terms of channels of distribution, it is obvious that convenience and service are key for the channels choice. These two factors are a proxy for trust. This result is consistent with the findings from the study conducted in Ontario [25], which also found a willingness to buy local food products if available in more conventional stores.

Although consistent with other research that has profiled a typical local food consumer, farmers should not solely target the typical demographic profile (well-educated woman with above average income and family) but should consider the importance of product attributes to all consumers when creating their marketing approach. For example, knowing that a product is locally produced, and promoting it based on quality indicators (e.g., nutrition, health benefits, taste, and reduced food mileage) might be a better strategy than just focusing on the typical local foods consumer. Contrary to the existing literature on sustainability, and the concept of embeddedness, this study did not indicate that the consumer's concerns and/or fears changed the consumer's decision to buy local. While the study does reveal that concerns have altered the purchasing patterns and behaviors of consumers, these concerns about foods might relate more to the Bovine Spongiform Encephalopathy (BSE) crisis for example than the fear of the globalized food system. Further exploration of the reasoning behind the decision to buy local could be explored in order to determine if social theory and the desire to purchase sustainable products plays a role in consumers' decision-making.

Author details

Mehdi Zahaf[2*] and Madiha Ferjani[1]

*Address all correspondence to: zahaf@telfer.uottawa.ca

1 Mediterranean School of Business, Tunisia

2 Telfer School of Management, University of Ottawa, Canada

References

[1] Aguirre, J.A. (2007), "The Farmer's Market Organic Consumer of Costa Rica", British Food Journal, Vol. 109 (2), 145-154.

[2] Aprile, M.C., Caputo, V., and Nayga, R.M. (2012), "Consumers' Valuation of Food Quality Labels: The case of the European Geographic Indication and Organic Farming Labels", International Journal of Consumer Studies, Vol. 36 (2), 158-165.

[3] Barham, E., Lind, D., and Jett, L. (2005), "The Missouri regional cuisines project: Connecting to place in the restaurant", In Barlett, F.P. (eds) Urban Place: Reconnections with the Natural World. MIT Press: Cambridge, MA.

[4] Bean, M., and Sharp, J. (2011), "Profiling alternative food system supporters: The personal and social basis of local and organic food support", Renewable Agriculture and Food Systems, Vol. 26, 243-254.

[5] Bérard, L., and Marchenay, P. (2004), "Les Produits de Terroir; Entre Cultures et Règlements. CNRS Editions, Paris.

[6] Botonaki, A., Polymeros, K., Tsakiridou, E., and Mattas, K. (2006), "The role of food quality certification on consumers' food choices", British Food Journal, Vol. 108 (2), 77-90.

[7] Browne, A.W., Harris, P.J.C., Hofny-Collins, A.H., Pasiecznik, N., and Wallace R.R. (2000), "Organic production and ethical trade: Definition, practice and links", Food Policy, Vol. 25, 69-89.

[8] Cunningham, R. (2001), "The organic consumer profile: Not only who you think it is!", Alberta Agriculture Food and Rural Development: Edmonton Alberta.

[9] Darby, K.B. (2008), "Decomposing Local: A Conjoint Analysis of Locally Produced Foods", American Journal of Agricultural Economics, 476-486.

[10] FiBL & IFOAM, 2015, "The world of organic agriculture", Frick and Bonn.

[11] Food Processing Centre (2001), "Attracting consumers with locally grown products", Institute of Agriculture and Natual Resources, University of Nebraska – Lincoln.

[12] Forrest, R. (2007), "Can you become a localvore?", Slow Food Vancouver, retrieved from http://www.slowfoodvancouver.com/index.php/Features/more/can_you_become_a_local.

[13] Fotopoulos, C., and Krystallis, A. (2002a), "Purchasing motives and profile of the Greek organic consumer: a countrywide survey", British Food Journal, Vol. 104 (9), 730-764.

[14] Fotopoulos, C., and Krystallis, A. (2002b), "Organic product avoidance: Reasons for rejection and potential buyers' identification in a countrywide survey", British Food Journal, Vol. 104, 233-260.

[15] Giovannucci, D., Barham, E., and Pirog, R. (2009) "Defining and marketing local foods: Geographical Indications for U.S. products", Journal of World Intellectual Property, special issue on GIs.

[16] Green, J.M., Draper, A.K., Dowler, E.A., Fele G., Hagenhoff, V., Rusanen, M., and Rusanen, T. (2005), "Public Understanding of Food Risks in Four European Countries: A Qualitative Study", European Journal of Public Health, Vol. 15 (5), 523-527.

[17] Gurviez, P., and Korchia, M. (2002), "Proposition d'une echelle de mesure multidimensionnelle de la confiance dans la marque", Recherche et Applications en Marketing, Vol. 17 (3), 41-61.

[18] Hair, Joseph F., Black W.C., Barry Babin, J., Anderson, R.E., and Tatham, R.L. (2006), Multivariate Data Analysis (6th ed.), Upper Saddle River, N.J.: Pearson Education Inc.

[19] Hamzaoui, L., and Zahaf, M. (2008), "Profiling Organic Food Consumers: Motivations, Trust Orientations and Purchasing Behavior", Journal of International Business and Economics, Vol. 8 (2), 25-39.

[20] Holt, G., and Amilien V. (2007), "From local food to localized food", Anthropology of Food, Special Issue on local food products and systems. Retrieved from https://aof.revues.org/405.

[21] Honkanen, P., Verplanken, B., and Olsen S.O. (2006), "Ethical Values and Motives Driving Organic Food Choice", Journal of Consumer Behavior, Vol. 5 (5), 420-430.

[22] Kristallis, A., and Chryssohoidis, G. (2005), "Consumers' willingness to pay for organic food: Factors that affect it and variation per organic product type", British Food Journal, Vol. 107 (5), 320-343.

[23] Larue, B., West, G., Gendron, C., and Lambert, R. (2004), "Consumer response to functional foods produced by conventional, organic, or genetic manipulation", Agribusiness, Vol. 20 (2), 155-166.

[24] Magkos, F., Arvanitif, F., and Zampelas, A. (2006), "Organic food: Buying more safety or just peace of mind? A critical review of the literature", Critical Review in Food Science and Nutrition, Vol. 46 No. 1, 23-56.

[25] Northwest Link (2009), Retrieved from http://www.ontariosoilcrop.org/docs/01NWLApr._09.pdf.

[26] Padel, S., and Foster, C. (2005), "Exploring the gap between attitudes and behaviour: Understanding why consumers buy or do not buy organic food", British Food Journal, Vol. 107 (8), 606-625.

[27] Pirog, R., and Larson, A. (2007), "Consumer Perceptions Of The Safety, Health, And Environmental Impact Of Various Scales And Geographic Origin Of Food Supply Chains", Leopold Centre.

[28] Raynolds (2004), "The Globalization of Organic Agro-food Networks, World Development, Vol. 32 (5), 725-743.

[29] Schwartz, S.H. (1977), "Normative Influences on Altruism", Advances in Experimental Social Psychology, (10), 221-279.

[30] Scialabba, N. (2000), "Factors Influencing Organic Agriculture Policies With A Focus On Developing Countries", IFOAM 2000 Scientific Conference, Switzerland.

[31] Sheperd, R., Magnusson, M., and Sjoden, P.O. (2005), "Determinants Of Consumer Behaviour Related To Organic Foods", Ambio, Vol. 34 (4/5), 352-359.

[32] Sirieix, L., Pontier, S., and Schaer, B. (2004), "Orientations De La Confiance Et Choix Du Circuit De Distribution: Le Cas Des Produits Biologiques", Proceedings of the 10th FMA International Congress, St. Malo, France.

[33] Soil Association (SA), 1999-2004, "Soil Association Organic Food and Farming Report", SA, Bristol.

[34] Tarkiainen, A., and Sundqvist, S. (2005), "Subjective Norms, Attitudes And Intentions Of Finnish Consumers In Buying Organic Food", British Food Journal, Vol. 107 No. 11, pp. 808-22.

[35] Urala, N., and Lahteenmaki, L. (2003), "Reasons Behind Consumers' Functional Food Choice", Nutrition and Food Science, Vol. 33 (4), 148-158.

[36] Van Elzakker B., and Eyhorn F. (2010), "Developing Sustainable Value Chains With Smallholders", The Organic Business Guide, Online Report.

[37] Weber, M. (1964), "Economy And Society", University of California Press. Retrieved from: https://docs.google.com/forms/d/16zP78RY0zQFemKdspwIz-CyEUIX24LHJc0ket_uQMpnE/viewanalytics.

[38] Wier, M., and Calverly, C. (2002), "Market Potential For Organic Foods In Europe", British Food Journal, Vol. 104 (1), 45-62.

[39] Wilkins, J.L., Bokaer-Smith, J., and Hilchey, D. (1996). "Local Foods And Local Agriculture: Survey Of Attitudes Among Northeastern Consumers", Northeast Regional Food Guide Project. Cornell.

[40] Willer, H., and Yussefi (2007), "The World Of Organic Agriculture", Online Report.

[41] Willer, H., and Kilcher, L. (2011), "The World of Organic Agriculture Statistics and Emerging Trends 2011", IFOAM, Bonn, & FiBL.

[42] Zanoli, R., and Naspetti, S. (2002), "Consumer Motivations in the Purchase of Organic Food: A Means End Approach", British Food Journal, Vol. 104, 643-653.

3

Organic Farming as an Essential Tool of the Multifunctional Agriculture

Elpiniki Skoufogianni, Alexandra Solomou, Aikaterini Molla and Konstantinos Martinos

Additional information is available at the end of the chapter

Abstract

This chapter aims at shedding light on the annals of organic farming and at defining its past and present meaning. Low-profile attempts were made in the first half of the last century when it comes to organic farming as it developed almost independently in the German and English speaking world. Organic farming has been established as a promising and innovative method of meeting agricultural needs and food production with respect to sustainability (climate change, food security and safety, biodiversity, rural development). Its value in terms of environmental benefits is also acknowledged. The differences between organic and conventional food stem directly from the farming methods that were used during the food items' production. Many people are unaware of some of the differences between the two practices. Agriculture has a direct effect on our environment, so understanding what goes into it is important. There are serious differences between organic and conventional farming; one of the biggest differences that is observed very frequently across all research between the two farming practices is the effect on the land. Conclusively, organic farming is a form of agriculture that relies on ecosystem management and attempts to reduce or eliminate external agricultural inputs, especially synthetic ones. It is a holistic production management system that promotes and enhances agro-ecosystem health, including biodiversity, biological cycles, and soil biological activity.

Keywords: Sustainability, environment, health, fertility

1. Introduction: History of organic farming

1.1. Growth and spread of the organic ideals

Many agricultural dogmas claim to strive towards sustainability [1]. Organic farming is the pinnacle of these models, and probably the one that is most acknowledged worldwide in the scientific and political arenas [2, 3], as well as by consumers as a whole. Today, organic farming is a legitimate system due to its history and evolution of practices, and rules and regulations [4, 5, 6, 7].

Organic farming is "a form of agriculture that uses fertilizers and pesticides (which include herbicides, insecticides and fungicides) if they are considered natural (such as bone meal from animals), but it excludes or strictly limits the use of various methods, including synthetic petrochemical fertilizers and pesticides; plant growth regulators such as hormones; antibiotic use in livestock; genetically modified organisms etc." [8]. As a result, it relies on techniques such as crop rotation, green manure, compost, and biological pest control.

Organic farming has dramatically grown in importance and influence worldwide throughout the years. A few statistics tell a fragment of the story: from almost negligible levels during the 1980s, the area of organic farms worldwide spanned to an estimated 43.1 million hectares in 2013 [9]; the worldwide organic market size was worth 54 billion euros in the same year [10]. However, these numbers depict only a part of what organic farming has become; scientists, educators, and agricultural policy makers have been making a change that formally began during the late 1970s. The growth of research on organic farming has been particularly striking, and the number and variety of organic curricula and degrees offered at universities in many countries are vast. At the first International Scientific Conference of the International Federation of Organic Agriculture Movements (IFOAM), held in Switzerland in 1977, a total of 25 presentations were offered. When the IFOAM conference returned to Switzerland in 2000, that number had multiplied more than 20 times, to well over 500 [11]. Before the 1970s, funds for organic research were extremely limited; today, significant public money is available in many countries: Denmark, France, Germany, Sweden, Switzerland, and the Netherlands are all reported to spend millions per year on organic research [12]. An important component of the advancement of organic farming has been its global spread. Five countries were represented when IFOAM was organized in 1972, and by the late 1990s, it had members from over 100 countries. IFOAM's scientific conferences, which until the mid-1980s had only been held in Western Europe and North America, have since been held in countries as diverse and dispersed as Burkina Faso, Australia, Hungary, and Brazil, among others. Further evidence that organic farming has gone global is that the UN Food and Agriculture Organization has been involved in it since 1999, with activities that include market analysis, environmental impact assessments, improving technical knowledge, and development of standards through the Codex Alimentarius Commission [13]. The United Nations Conference on Trade and Development has been involved in global trade of organic foods since 2001, particularly in assisting developing countries in increasing their production [14].

1.2. History

The concept we know today as 'organic farming' is a mixture of different views coming mainly from German and English-speaking societies. These ideas arose at the end of the 19th century, and between the two World Wars, as intensive and mechanized farming faced a crisis in the form of soil degradation, poor food quality and the decay of rural social life and traditions.

Inappropriate use of mineral fertilizers was disturbing plant metabolism, making them susceptible to pathogens and insect pests. At the same time, effective pesticides had not yet been developed. Physiologically acidic mineral fertilizers acidified the soil and brought about diminished root growth and degradation of the soil structure. Soil compaction caused by the use of machinery and reduced organic manuring caused droughts, and soils experienced a decline in fertility – referred to as "soil fatigue" (Bodenmüdigkeit) [15]. Despite the increased use of mineral fertilizers, agriculture suffered a dramatic drop in yields (up to 40% in countries like Germany) after World War I; only at the end of the 1930s – after more than 15 years – did yields reach pre-war levels [16].

Some consumers were worried about declining food quality: food that did not stay fresh, tasteless vegetables and fruits, and pesticide residue based on toxic heavy metals. Increased use of mineral fertilizers and pesticides was discussed by the public as a major cause of this decline. For example, an assumption that an elevated level of potassium in cancer cells was caused by the increased amount of potassium in fertilizers was not something unthought-of. Scientists such as Robert McCarrison in the UK or Werner Schuphan and Johannes Görbing in Germany confirmed some of these suspicions, such as lower vitamin levels in fruits and vegetables caused by increased nitrogen fertilization [17, 18]. Finally, the social and economic situation in the countryside changed dramatically with the mechanization of agriculture, industrialization of the food sector, and import of agricultural products. An imbalance arose between the urban centers; severe economic problems caused by low prices (due to imports) and indebtedness (due to purchase of machines, fertilizers, and pesticides) forced many small and medium-sized farms to give up. As a result, there was a general decline in rural tradition and lifestyle.

As a solution to this crisis, organic farming pioneers offered a convincing, science-based theory during the 1920s and 1930s that evolved into a successful farming system during the 1930s and 1940s. But it was not until the 1970s, with growing awareness of an environmental crisis, that organic farming attracted interest in the wider worlds of agriculture, society, and politics. The leading strategies that proposed to achieve sustainable land use included a biological concept of soil fertility, intensification of farming by biological and ecological innovations, renunciation of artificial fertilizers and synthetic pesticides to improve food quality and the environment and, finally, concepts of appropriate animal husbandry.

At the annual meeting of the American Association for the Advancement of Science (AAAS) in 1974, a panel of scientists targeted the "organic food myth", calling it "scientific nonsense" and the domain of "food faddists and eccentrics". They also blamed such "pseudoscientists" for causing panic among the public with regard to paying more for food [19] and also mentioned that the "organic myth" was counterproductive to human welfare, because it leads to a rejection of procedures that are needed for the production of nutritious food at "maximum efficiency" and was "eroding gains of decades of farming advancements". However, 7 years later, the journal of this same AAAS published a major research paper that found organic farms to be highly efficient and economically competitive when compared to conventional farming [7].

2. Comparison of organic and conventional farming system

In the recent years, agriculture has been oriented towards industrial and notably intensive farming practices aimed at ensuring enough food for humanity. However, these types of farming practices also caused several negative environmental impacts such as decreasing biodiversity. Many agroecosystems intensified their activities and became highly mechanized, while those unable to do so became increasingly marginalized and were sometimes forced to abandon their land, causing evenly destructive effects for biodiversity [20].

Currently, it is globally imperative that the increasing demand for food be met in a manner that is socially fair and ecologically sustainable over the long run. It is possible to design farming systems that are similarly productive and that enhance the provisioning of ecosystem services such as biodiversity, soil quality and nutrient, control of weeds, diseases and pests, energy efficiency, and the reduction of global warming potential, as well as resistance and resilience to climate change and crop productivity [21].

Organic farming is a system that favors soil fertility by maximizing the efficient use of local resources, while foregoing the use of agrochemicals and genetically modified organisms. The high quality of organic food and its added value based on a number of farming practices relies on ecological cycles, and it focuses on declining the environmental effect of the food industry, maintaining long-term sustainability of soil and reducing to a minimum the use of nonrenewable resources [22].

Organic farming practices have been launched to reduce the environmental impacts of agriculture. The results of studies that compare the environmental impacts of organic and conventional farming in Europe show that organic farming has a positive impact on the environment. Important differences between the two farming systems include soil organic matter (SOM) content, nitrogen leaching, nitrous oxide emissions, energy use, and land use. Most of the studies that compared biodiversity in organic and conventional farming showed lower environmental impacts from organic farming [23].

Furthermore, organic farming appears to perform better than conventional farming and also provides other important environmental advantages such as curbing the use of harmful chemicals and their spread in the environment and along the trophic chain, and reducing water use [22].

• *Health*

Organic practices contribute to better health through reduced pesticide exposure for all and increased nutritional quality in food products. In order to understand the importance of consumption of organic food from the viewpoint of toxic pesticide contamination, we should look at the whole picture: from the farmers who do the valuable work of growing food, to the waterways from which we drink, the air we breathe, and the food we eat. Organic food can nourish us and keep us healthy without causing the toxic effects of chemical agriculture [24, 25].

The population groups most affected by pesticide use are farmers. These people live in communities near the application of toxic pesticides, where pesticide drift and water contamination are common. Farmers, both pesticide applicators and fieldworkers who tend to and harvest the crops, come into frequent contact with such pesticides. Organic farming does not utilize these toxic chemicals, and thus eliminates this enormous health hazard to workers, their families, and their communities [25, 26].

Acute pesticide poisoning among farmers is only one aspect of the health consequences of pesticide exposure. Many farmers spend time in the fields, resulting in prolonged exposure, and some studies have reported increased risks of certain types of cancers among farmers as a consequence. The emerging science on endocrine disrupting pesticides reveals another chronic health effect of pesticide exposure [25, 27].

• *Environment*

Organic farming is often perceived to have generally beneficial effects on the environment compared to conventional farming [28, 29]. More specifically, organic food production eliminates soil and water contamination. Since organic food production strictly avoids the use of all-synthetic chemicals, it does not pose any risk of soil and underground water contamination like conventional farming, which uses tons of artificial fertilizers and pesticides. Also, organic food production helps preserve local wildlife; by avoiding toxic chemicals, using mixed planting as a natural pest control measure, and maintaining field margins and hedges, organic farming provides a retreat to local wildlife rather than taking away their natural habitat like conventional agriculture [30].

Agrobiodiversity is an important aspect of biodiversity that is directly influenced by different production methods, especially at the field level. It can also supply several ecosystem services to agriculture, thus reducing environmental externalities and the need for off-farm inputs. Moreover, organic farming helps conserve biodiversity. Avoidance of chemicals and use of alternative, all natural farming methods have been shown to help conserve biodiversity as it encourages a natural balance within the ecosystem and helps prevent the domination of a particular species over the others [31].

Various different approaches have been used in order to compare environmental impacts of farming systems, such as organic and conventional. Several studies have focused on biodiversity [31, 32], land use [33], soil properties [34, 35], or nutrient emissions [36, 37]. Life cycle assessment (LCA) studies have used a product approach to assess the environmental impacts of a product from input production up to the farm gate [38, 39]. According to the literature, Mondelaers et al. (2009) [40] used the meta-analysis method to compare the environmental impacts of organic and conventional farming, examining land-use efficiency, organic matter content in the soil, nitro-phosphate leaching into the water system, greenhouse gas (GHG) emissions, and the effect on biodiversity [23].

In a review of literature, Hole et al. (2005) [31] compared biodiversity in organic and conventional agroecosystems. They found that organic farming generally had positive impacts on biodiversity. However, they concluded that it is still unclear whether conventional farming with specific practices for biodiversity conservation (i.e., agri-environmental schemes) can provide higher benefits than organic farming. More studies published after 2003 supported the findings of Hole et al. (2005) [31] and Bengtsson et al. (2005) [41], but none found organic farming to have negative impacts on biodiversity. More specifically, herbaceous plant richness has been widely found to be higher in organic farms compared with conventional farms [42, 43], and several studies showed that landscape had more important impact on biodiversity than farming practices [44, 45]. It has also been found that organic farming, without additional practices, is not adequate for conserving some animal species [23, 44, 46, 48].

The main reason for the reduction of agricultural biodiversity during the last decades has been the change in agricultural landscapes [48, 49]. In Europe, formerly heterogeneous landscapes with a mix of small arable agroecosystems, semi-natural grasslands, wetlands, and hedgerows have been replaced in many areas by largely homogeneous areas of intensively cultivated farms [50]. This has resulted in declines in biodiversity and has caused an important loss of species [23, 51].

Regarding the soil ecosystem, Tuomisto et al. (2012) [23] had found that organic matter across all the cases was 7% higher in organic farms compared to conventional farms. The main explanation for higher organic matter contents in organic systems was that they had higher organic inputs such as manner or compost. Other explanations for higher SOM levels in organic systems were less intensive tillage and inclusion of leys in the rotation [52, 53]. Gosling and Shepherd (2005) [54] observed lower organic matter contents in organic farms by higher yields, and thus, higher crop residue leftovers in conventional systems, which can compensate the lower external organic matter inputs. Furthermore, they argued that leys do not necessarily contribute to the increase of organic matter because they have a low carbon–nitrogen ratio and, therefore, organic matter decomposes quickly.

According to some studies [55, 56], the main explanation for lower nitrogen leaching levels from organic farming per unit of area was the lower levels of nitrogen inputs applied. Raised nitrogen leaching levels were explained by bad synchrony between the nutrient availability and crops' nutrient intake [57]. Notably, after incorporation of leys, the nitrogen losses tend to be high [58].

In conclusion, organic farming is a method of crop and livestock production that considered an environmentally friendly agriculture practice and a holistic approach involving several requirements and prohibitions from a regulatory point of view, and receives primarily from European countries additional agri-environmental payments for ecosystem services such as biodiversity. In several countries, payments are available as single biodiversity measures such as insectary strips, hedgerows, crop rotation, or the retention of semi-natural areas in agri-environmental programs that also focus on conventional farming.

3. Organic farming, conservation agriculture, and sustainability

This chapter shows the connection between organic farming and sustainability-conservation models, how this interplay has evolved during the past years, and, more importantly, its future directions. Various agricultural models claim to achieve sustainability. Organic farming is one of those candidate models, and probably the most widely known and accepted on an international level. It is recognized in the scientific and political areas as well as by society as a whole. Organic farming has been established as a promising and innovative method of meeting agricultural needs and food production with respect to sustainability (climate change, food security and safety, biodiversity, rural development). Its value in terms of environmental benefits is also acknowledged.

Organic agriculture is developing rapidly, and statistical information is now available from 138 countries in the world. Its share of agricultural land and farms continues to grow in many countries. According to the latest survey on organic farming worldwide, almost 30.4 million hectares are managed organically by more than 700,000 farmers. Most of this land is in Latin America, followed by Asia, Africa, and Europe [9].

Organic farming works in harmony with nature rather than against it, and it involves using techniques to achieve good crop yields, without harming the natural environment, or the people who live and work in it. The methods and materials that organic farmers use are summarized as follows:

To keep and build good soil structure and fertility:

- Recycled and composted crop wastes and animal manures
- Right soil cultivation at the right time
- Crop rotation
- Green manures and legumes
- Mulching on the soil surface

To control pests, diseases, and weeds:

- Careful planning and crop choice
- The use of resistant crops
- Good cultivation practice
- Crop rotation
- Encouraging useful predators that eat pests
- Increasing genetic diversity
- Using natural pesticides

Organic farming also involves:

- Careful use of water resources
- Good animal husbandry

Future global food security relies not only on high production and access to food but also on the need to address the destructive effects of current agricultural production systems on ecosystem services [65] and to increase the resilience of the production systems to the effects of climate change. Conservation agriculture (CA) enables the sustainable intensification of agriculture by conserving and enhancing the quality of the soil, leading to higher yields and the protection of the local environment and ecosystem services [67].

CA is a concept for resource-saving agricultural crop production that strives to achieve acceptable profits together with high and sustained production levels, while concurrently conserving the environment. CA is based on enhancing natural biological processes above and below the ground. Interventions such as mechanical soil tillage are reduced to an absolute minimum, and the use of external inputs such as agrochemicals and nutrients of mineral or organic origin are applied at an optimum level and in a way and quantity that does not interfere with, or disrupt, the biological processes.

CA is characterized by three principles which are linked to each other, namely:

1. Continuous minimum mechanical soil disturbance (i.e., no tilling and direct planting of crop seeds).

2. Permanent organic soil cover.

3. Diversification of crop species grown in sequence and associations [62].

It has generally been demonstrated that CA allows yields to increase while improving soil and water conservation, and reducing production costs [60, 64]. In addition, CA has been shown to work successfully in a variety of agroecological zones and farm sizes. Indeed, another advantage associated with CA is that it can be applied to different farming systems, with different combinations of crops, sources of power and production inputs.

There is no real dispute that sustainable agriculture and organic farming are closely related terms. There is, however, some disagreement on the exact nature of this relationship; for some, the two are synonymous, while for others, equating them is misleading. Lampkin's definition of organic farming, quoted earlier, talks of sustainable production systems. Having provided his definition, he goes on to state: "...sustainability lies at the heart of organic farming and is one of the major factors determining the acceptability or otherwise of specific production practices." Similarly, Henning et al. precede their definition of organic farming, quoted above, by claiming that "it could serve equally well as a definition of 'sustainable agriculture'" [59]. Rodale even suggested that "sustainable was just a polite word for organic farming" [63]. Some of the research that has been carried out regarding the historical relationship between agricultural systems and the sustainability of the societies they support illustrates the point that a farming system need not be modern, mechanized, and using synthetic chemicals to be profoundly unsustainable [61].

Part of the difficulty in assessing the sustainability of agricultural systems, is the fact that both the units of measurement and the appropriate scales for measurement differ both within and across the commonly identified economic, biophysical and social dimensions of sustainability. For example, consideration of the effects of organic production on farm margins, soil fertility, and rural employment are difficult to combine in an overall measure. They are not so problematic if the effects are all in the same direction, but when one starts to consider trade-offs, as one indicator increases and another falls across different dimensions, then this factor becomes more significant. This is an issue which will not be solved simply by greater knowledge of the impacts of different production systems; even with complete information regarding impacts, one will still have to consider trade-offs with movement towards targets in some respects accompanied by reverses in others [61].

4. Organic practices

Throughout the years, organic farming has evolved in a diverse manner. Many sub-schools and sub-dogmas have appeared. Two of the most important, biointesive farming and permaculture, are discussed below:

4.1. Biointesive farming

Biointensive agriculture aims to result in maximum yields from the minimum area of land, while simultaneously improving and maintaining the fertility of the soil, as well as abiding by the rules of organic farming all the time. It is particularly designed for the small-scale grower. Biointensive cropping strategies (i.e., polycultures) are usually labor intensive [68].

4.1.1. Permaculture

Permaculture emphasizes eco-design [69]. Sepp [70] defines permaculture as a system in which every element fulfills multiple functions, and every function is performed by multiple elements. Energy is used practically and efficiently with a great focus on renewable forms, and diversity is favored instead of monoculture.

4.2. Crop rotation

Crop rotation is a very important piece of all organic cropping systems because it provides the basic function of keeping soils healthy, an efficient way to control pests, and other benefits. Crop rotation is defined as changing the type of crop grown on a particular piece of land from year to year [71]. There are both cyclical rotations, in which the same sequence of crops is repeated on the same field, and noncyclical rotations, in which the sequence of crops is diversified to meet the changing needs of the farmer.

Good crop rotation requires long-term strategic planning. However, planning that is too long term may prove futile as choices can be affected by changes in weather, in the market, labor expenses, and other factors. Conversely, lack of planning can lead to serious problems – for example, the buildup of soil-borne diseases of a critical crop, or imbalances in nutrients [71]. Problems like the ones mentioned above often take several years to become noticed and can catch even experienced farmers by surprise. In fact, rotation problems usually do not develop until well after the transition to organic cropping. Fallowing is also a noted part of crop rotation.

The design of a diverse crop rotation is the key to soil nutrients, weed, pest, and disease management. To achieve even some of these benefits of crop rotation, great focus on management is required, since diversity simply as a goal may lead to losses in production and productivity [72]. Therefore, there is a need for functional diversity [73]. In mixed inter-cropping, crop cycles tend to be similar to allow simultaneous management of the components (e.g., grass/clover leys or cereal with grain legumes), or completely different to allow separate management (e.g., cereals intercropped with forage legumes). Extremes of mixed intercrop-ping systems can be seen in agroforestry [74] or perennial polyculture [75, 76].

Principles guiding the spatial arrangement of crops in polyculture are also well developed, dominantly originating from horticulture; they have been tested through research and developed by trial and error [77, 78, 79] of studies of traditional cropping systems [80, 81, 82].

4.3. No till and conservation till farming

In zero tillage, the soil is left undisturbed from harvest to planting, except for nutrient supply. Planting or drilling is accomplished in a narrow seedbed or slot created by coulters, row cleaners, disk openers, as well as in-row chisels [83]. Weed control is accomplished primarily with herbicides.

Conservation tillage is defined as tillage and planting system that maintains at least 30% of the soil surface covered by residue after planting (CTIC and Conservation Technology Information 1998). There are various benefits to this practice, with the most important being economic (conservation tillage operations reduce costs) and environmental (reduced cultiva-tion implies reduced energy inputs [84], thereby ensuring less pollution and less disturbed soil, while organic matter accumulation is increased and CO_2 releases to the atmosphere are much reduced [85]).

4.4. Mulching

Mulching is the method of covering the surface of the soil with any decomposable material (grass, hay, leaves, waste etc.) Benefits include the soil is not dried by wind and sun exposure, moisture is reserved and soil erosion is prevented, rich humus is provided to the soil, and soil drainage is improved. It also leads to an increase in soil micro organisms and reduction in weed growth.

4.5. Composting

Composting is a process where microorganisms decompose organic matter to produce a humus-like substance called compost. The process is natural, provided the right organisms, water, oxygen, organic material, and nutrients are in place. By controlling these factors, the composting process can occur at a much faster rate [86]. The bacteria and fungi occurring in the soil convert dead organic matter present on its surface into a nutrient-rich medium. This is called composting, and the nutrient-rich medium is called compost. Following are the benefits of compost, compared to the usage of raw manure:

1.   Making compost turns waste into a profitable resource.

2.   Compost is environmentally friendly and promotes industry sustainability.

3. Compost adds organic material, thereby improving the soil structure and water retention.

4. Compost use reduces the need for inorganic fertilizers.

5. Causes slow release of nutrients – nutrients are released to the plants slowly, thus reducing the loss of nutrients to the environment.

Author details

Elpiniki Skoufogianni[1], Alexandra Solomou[2*], Aikaterini Molla[3] and Konstantinos Martinos[1]

*Address all correspondence to: alexansolomou@gmail.com

1 Laboratory of Agronomy and Applied Crop Physiology, Dept. of Agriculture, Crop Production and Rural Environment, University of Thessaly, Volos, Greece

2 National Agricultural Research Foundation, Institute of Mediterranean Forest Ecosystems Terma Alkmanos, Ilisia, Athens, Greece

3 National Agricultural Research Foundation, Larisa, Greece

References

[1] Koohafkan P, Altieri MA, Gimenez EH. Green agriculture: Foundations for biodiverse, resilient and productive agricultural systems. International Journal of Agricultural Sustainability. 2011;10: 61–75.

[2] McIntyre BD, Herren HR, Wakhungu J, Watson RT. International Assessment of Agricultural Knowledge, Science and Technology for Development (IAASTD): Global Report. Island Press, Washington, DC, 2009.

[3] National Research Council. Toward sustainable agricultural systems in the 21st century. The National Academies, Washington, DC, 2010.

[4] Besson Y. Une histoire d'exigences: philosophie et agrobiologie. L'actualité de la pensée des fondateurs de l'agriculture biologique pour son développement contemporain. Innovations Agronomiques, 2009;4: 329–362.

[5] Francis C editor. Organic Farming: The Ecological System. American Society of Agronomy, Inc., Crop Science Society of America, Inc., Soil Science Society of America, Inc.: Madison, 2009; 353 p.

[6] Kristiansen P, Taji A, Reganold J. Organic agriculture: opportunities and challenges, in: PTARJ Kristiansen (ed.), Organic agriculture: a global perspective, Cabi, Wallingford, 2006.

[7]   Lockeretz, W. 'What explains the rise of organic farming', in W. Lockeretz (ed.), Organic Farming: An International History. Wallingford, CABI, 2007.

[8]   European Commission official website, 2014.

[9]   Lernoud W, Lernoud H, Lernoud J (eds), The World of Organic Agriculture. Statistics and Emerging Trends 2015. FiBL-IFOAM Report. Bonn, 2015.

[10] Organics international, consolidated annual report of IFOAM - Organics International. Bonn, 2014.

[11] Alföldi T, Lockeretz W and Niggli U (eds), IFOAM 2000 – The World Grows Organic. Proceedings of the 13th International IFOAM Scientific Conference, Basel, 28– 31 August, 2000.

[12] Slabe A. Consolidated Report: Second Seminar on Organic Food and Farming Research in Europe: How to Improve Trans-national Co- operation, 2004. Available from: http://www. agronavigator.cz/attachments/CORE_ seminar_listopad_2004.pdf

[13] FAO. Organic agriculture at FAO. United Nations Food and Agriculture Organization. 2005. Available from: www.fao.org/organicag.

[14] Twarog S. UNCTAD's work on organic agriculture. In: Rundgren, G. and Lockeretz, W. (eds) Reader, IFOAM Conference on Organic Guarantee Systems: International Harmonization and Equivalence in Organic Agriculture, 17–19 February 2002, Nuremberg, Germany. IFOAM, Tholey-Theley, Germany, 2002.

[15] Vogt G. Entstehung und Entwicklung des ökologischen Landbaus im deutschsprachigen Raum. Bad Dürkheim: SÖL, 2000.

[16] Bittermann E. Die landwirtschaftli- che Produktion in Deutschland 1800–1950. Kühn-Archiv. 1956; 70: 1–145.

[17] McCarrison R and Viswanath, B. The effect of manural conditions on the nutritive and vitamin values of millet and wheat. Indian Journal of Medical Research. 1926; 14: 351–378.

[18] Schuphan W. Untersuchungen über wichtige Qualitätsfehler des Knollenselleries bei gleichzeitiger Berücksichtigung der Veränderung wertgebender Stoffgruppen durch die Düngung. Bodenkunde und Pflanzenernährung. 1937; 2: 255–304.

[19] Washington Post. Organic farming "scientific nonsense". 1974.

[20] Lockeretz W, Shearer G and Kohl D. Organic Farming in the Corn Belt. Science. 1981; 211: 540–547.

[21] European Commission. The EU Biodiversity Strategy to 2020 [internet]. 2011. Available from: http://ec.euro.eu/environment/nature/info/pubs/docs/brochures/ 2020%20Biod%20brochure%20final%20lowres.pdf.

[22] Kremen C, Miles A. Ecosystem services in biologically diversified versus conventional farming systems: benefits, externalities, and trade-offs. Ecology and Society. 2012; 17: 40. http://dx.doi.org/10.5751/ES-05035-170440.

[23] Gomiero T, Pimentel D, Paoletti MG. Environmental Impact of Different Agricultural Management Practices: Conventional vs. Organic Agriculture. Critical Reviews in Plant Sciences. 2011; 30: 95-124.

[24] Tuomisto HL, Hodge ID, Riordan P, Macdonald DW. Does organic farming reduce environmental impacts? - A meta-analysis of European research. Journal of Environmental Management. 2012; 112: 309-320.

[25] Givens I, Baxter S, Minihane AM, Shaw E. Health Benefits of Organic Food Effects of the Environment. Cromwell Press, Trowbridge. 2008; 315 p.

[26] http://beyondpesticides.org/organicfood/health/index.php

[27] Reeves M, Katten A, Guzmán M. Fields of Poison, California Farmworkers and Pesticides. Reports by Californians for Pesticide Reform. 2002; 37 p.

[28] FIAN. Pestizide – Eine Gefahr für die Umsetzung des Rechts auf Nahrung, Münster; 2011.

[29] Aldanondo-Ochoa AM, Almansa-Saez C. The private provision of public environment: consumer preferences for organic production systems. Land Use Policy. 2009; 26: 669-682.

[30] Gracia A, de Magistris T. The demand for organic foods in the South of Italy: A discrete choice model. Food Policy. 2008; 33: 386-396.

[31] Letourneau DK, Bothwell SG. Comparison of organic and conventional farms: challenging ecologists to make biodiversity functional. Frontiers in Ecology and the Environment. 2008; 6: 430–438.

[32] Hole DG, Perkins AJ, Wilson JD, Alexander IH, Grice PV, Evans AD, Does organic farming benefit biodiversity? Biological Conservation. 2005; 122: 113–130.

[33] Rundlof M, Nilsson H, Smith HG. Interacting effects of farming practice and landscape context on bumblebees. Biological Conservation. 2008; 141: 417-426.

[34] Badgley C, Moghtader J, Quintero E, Zakem E, Chappell MJ. Aviles-Vazquez, K., Samulon, A., Perfecto, I., Organic agriculture and the global food supply. Renewable Agriculture and Food Systems 2007; 22: 86-108.

[35] Maeder P, Fliessbach A, Dubois D, Gunst L, Fried P, Niggli U. Soil fertility and biodiversity in organic farming. Science. 2002; 296: 1694-1697.

[36] Stockdale EA, Shepherd MA, Fortune S, Cuttle SP. Soil fertility in organic farming systems - fundamentally different? Soil Use and Management 2002; 18: 301-308.

[37] Syvasalo E, Regina K, Turtola E, Lemola R, Esala M. Fluxes of nitrous oxide and methane, and nitrogen leaching from organically and conventionally cultivated san-

dy soil in western Finland. Agriculture, Ecosystems & Environment. 2006; 113: 342-348.

[38] Trydeman Knudsen M, Sillebak Kristensen IB, Berntsen J, Molt Petersen B, Steen Kristensen E. Estimated N leaching losses for organic and conventional farming in Denmark. The Journal of Agricultural Science. 2006; 144: 135-149.

[39] Cederberg C, Mattsson B. Life cycle assessment of milk production- a comparison of conventional and organic farming. Journal of Cleaner Production. 2000; 8: 49-60.

[40] Thomassen MA, van Calker KJ, Smits MCJ, Iepema GL. de Boer, I.J.M., Life cycle assessment of conventional and organic milk production in the Netherlands. Agricultural Systems. 2008; 96: 95-107.

[41] Mondelaers K, Aertsens J, Van Huylenbroeck G. A meta-analysis of the differences in environmental impacts between organic and conventional farming. British Food Journal. 2009; 111: 1098-1119.

[42] Bengtsson J, Ahnstrom J, Weibull AC. The effects of organic agriculture on biodiversity and abundance: a meta-analysis. Journal of Applied Ecology. 2005; 42: 261-269.

[43] Gabriel D, Roschewitz I, Tscharntke T. Thies, C., Beta diversity at different spatial scales: plant communities in organic and conventional agriculture. Ecological Applications. 2006; 16: 2011-2021.

[44] Romero A, Chamorro L, Sans FX. Weed diversity in crop edges and inner fields of organic and conventional dryland winter cereal crops in NE Spain. Agriculture, Ecosystems & Environment. 2008; 124: 97-104.

[45] Piha M, Tiainen J, Holopainen J, Vepsalainen V. Effects of land-use and landscape characteristics on avian diversity and abundance in a boreal agricultural landscape with organic and conventional farms. Biological Conservation. 2007; 140: 50-61.

[46] Rundlof M, Nilsson H, Smith HG. Interacting effects of farming practice and landscape context on bumblebees. Biological Conservation. 2008; 141: 417-426.

[47] Kragten S, Snoo GRD. Nest success of Lapwings Vanellus vanellus on organic and conventional arable farms in the Netherlands. Ibis 2007; 149: 742-749.

[48] Ekroos J, Piha M, Tiainen J. Role of organic and conventional field boundaries on boreal bumblebees and butterflies. Agriculture, Ecosystems & Environment 2008; 124: 155-159.

[49] Luoto M, Pykala J, Kuussaari M. Decline of landscape-scale habitat and species diversity after the end of cattle grazing. Journal for Nature Conservation (Jena) 2003; 11: 171-178.

[50] Gabriel D, Thies C, Tscharntke T. Local diversity of arable weeds increases with landscape complexity. Perspectives in Plant Ecology Evolution and Systematics 2005; 7: 85-93.

[51] Benton TG, Vickery JA, Wilson JD. Farmland biodiversity: is habitat heterogeneity the key? Trends in Ecology & Evolution 2003; 18: 182-188.

[52] Krebs JR, Wilson JD, Bradbury RB, Siriwardena GM. The second silent spring? Nature. 1999; 400: 611-612.

[53] Quintern M, Joergensen RG, Wildhagen H. Permanent-soil monitoring sites for documentation of soil-fertility development after changing from conventional to organic farming. Journal of Plant Nutrition and Soil Science Zeitschrift Fur Pflanzenernahrung Und Bodenkunde. 2006; 169: 564-572.

[54] Canali S, Di Bartolomeo E, Trinchera, A, Nisini L, Tittarelli F, Intrigliolo F, Roccuzzo G, Calabretta ML. Effect of different management strategies on soil quality of citrus orchards in Southern Italy. Soil Use and Management. 2009; 25: 34-42.

[55] Gosling P, Shepherd M. Long-term changes in soil fertility in organic arable farming systems in England, with particular reference to phosphorus and potassium. Agriculture, Ecosystems & Environment. 2005; 105: 425-432.

[56] Korsaeth A. Relations between nitrogen leaching and food productivity in organic and conventional cropping systems in a long-term field study. Agriculture, Ecosystems & Environment. 2008; 127: 177-188.

[57] Torstensson G, Aronsson H, Bergstrom L. Nutrient use efficiencies and leaching of organic and conventional cropping systems in Sweden. Agronomy Journal. 2006; 98: 603-615.

[58] Aronsson H, Torstensson G, Bergstrom L. Leaching and crop uptake of N, P and K from organic and conventional cropping systems on a clay soil. Soil Use and Management. 2007; 23: 71-81.

[59] Syvasalo E, Regina K, Turtola E, Lemola R, Esala M. Fluxes of nitrous oxide and methane, and nitrogen leaching from organically and conventionally cultivated sandy soil in western Finland. Agriculture, Ecosystems & Environment. 2006; 113: 342-348.

[60] Henning J., Baker L., Thomassin P. Economic issues in organic agriculture Canadian Journal of Agricultural Economics. 1991; 39: 1991 877–889.

[61] Kassam A., Friedrich T., Shaxson, F., Pretty J. The spread of Conservation Agriculture: Justification, sustainability and uptake. International Journal of Agricultural Sustainability. Sustainability. 2009; 7(4): 292-320.

[62] Rigby D., Cáceres D. Organic farming and the sustainability of agricultural systems. Agricultural Systems. 2001; 68(1): 21–40.

[63] Silici L. Conservation Agriculture and Sustainable Crop Intensification in Lesotho. Integrated Crop Management. 2010; 9-10.

[64] York E.T. Jr.Agricultural sustainability and its implications to the horticulture profession and the ability to meet global food needs. HortScience. 1991; 26(10): 1252–1256.

[65] FAO, 2001b. The economics of conservation agriculture. FAO. Rome, Italy.

[66] Foresight. The Future of Food and Farming: Challenges and Choices for Global Sustainability. Final Project Report. The Government Office for Science, London, 2011.

[67] Friedrich T., Kassam A., Shaxson F. Conservation Agriculture (CA). Agricultural Technologies for Developing Countries, Annex 2. European Technology Assessment Group, FAO, Rome. 2008.

[68] Willer H, Yussefi M, Sorensen N. The world of organic agriculture: statistics and emerging trends 2008, 2010.

[69] Guthman, J. Agrarian Dreams. The paradox of organic farming in California. Berkeley, University of California Press. 2004.

[70] Mollison B, Holmgren D. Permaculture one: a perennial agriculture for human settlements. Transworld (Corgi, Bantam), Melbourne. 1978.

[71] Holzer S. Sepp Holzer's permaculture: A practical guide to small-scale, integrative farming and gardening, Chelsea Green Publishing White River Junction, Vermont, 2001.

[72] Charles LM and Johnson SE (eds), Crop Rotation on Organic Farms: A Planning Manual, NRAES 177, 2009.

[73] Altieri M A. The ecological role of biodiversity in agroecosystems. Agriculture Ecosystems & Environment. 1991; 74: 19–31.

[74] Stockdale EA, Lampkin NH, Hovi M, Keatinge R, Lennartsson EKM, Macdonald DW, Padel S, Tattersall FH, Wolfe MS, Watson CA. Agronomic and environmental implications of organic farming systems. Advances in Agronomy. 2001; 70: 261-327.

[75] Nair PK. An Introduction to Agroforestry. Kluwer Academic Publishers. Dordrecht. Neuerburg, W., and Padel, S. (1992). In "Organisch-Biologischer Landbau in der Praxis." BLVVerlag, München, 1993.

[76] Jackson W. In "New Roots for Agriculture." Friends of the Earth, San Francisco. Jansen, K. (1999). Labour, livelihoods and the quality of life in organic agriculture. Biological Agriculture and Horticulture 1980; 17: 247–278.

[77] Soule JD and Piper JK. In "Farming in Nature's Image: an Ecological Approach to Agriculture." Island Press, Washington D.C., 1992.

[78] Lockhart, JAR and Wiseman AJL. In "Introduction to Crop Husbandry Including Grass- land." Pergamon Press, Oxford, 1988.

[79] Finch S. Entomology of crucifers and agriculture—diversification of the agroecosystem in re- lation to pest damage in cruciferous crops. In "The Entomology of Indige-

nous and Naturalized Systems in Agriculture." M. K. Harris, and C. E. Rogers (eds), 1988; 39–71.

[80] Theunissen J. Intercropping in field vegetables as an approach to sustainable horti-culture. Outlook on Agriculture 1997; 26: 95–99.

[81] Francis CA. Introduction: Distribution and importance of multiple cropping. In "Multiple Cropping Systems." C. A. Francis (ed.), 1–19. Macmillan Publishing Company, New York, 1986.

[82] Liebman M and Dyck E. Crop rotation and intercropping strategies for weed management. Ecological Applications 1995; 92–122.

[83] Lockhart JAR, and Wiseman, AJL. In "Introduction to Crop Husbandry Including Grass- land." Pergamon Press, Oxford, 1988.

[84] Altieri MA, Nicholls CI, and Wolfe MS. Biodiversity—a central concept in organic agri- culture: Restraining pests and diseases. In "Fundamentals of Organic Agriculture. Vol. 1"; (T. V. Ostergaard, Ed.), pp. 91–112. IFOAM: Ökozentrum Imsbach, D-66636 Tholey-Theley,1996.

[85] Stonehouse DP, Weise, SF, Sheardown T, Gill RS, and Swanton CJ. A case study approach to comparing weed management strategies under alternative farming strategies in Ontario. Canadian Journal of Agricultural Economics. 1996; 44: 81–99.

[86] Halvorson AD, Wienhold BJ, Black AL. Tillage, nitrogen, and cropping effects on soil carbon sequestration. Soil Science Society of America Journal. 2002;66: 906–912.

[87] Shiva V, Pande P and Singh J. Principles of Organic Farming. Renewing the Earth's Harvest, Navdanya, New Delhi, 2004.

4

The Use of Organic Foods, Regional, Seasonal and Fresh Food in Public Caterings

Jan Moudry Jr, Jan Moudry and Zuzana Jelinkova

Additional information is available at the end of the chapter

Abstract

The chapter focuses on possibilities to improve the quality of meals in public, especially school catering facilities. It presents the options for diet modifications towards a sustainable use of organic foods, local and seasonal food by optimizing portions of meat and meals prepared of fresh ingredients. From an economic, environmental and nutritional point of view, evaluation and comparison of the original and optimized meals can contribute to a more efficient use of foods and motivate staff in public catering facilities to comprehensive food assessment.

An overall evaluation shows that more favourable nutritional parameters may be achieved by the optimization of meals. A greater use of local, seasonal and organic foods, a reduction in meat portions and lower level of processing make energy and greenhouse gas emission savings and it is possible to sustain the costs within standard. The purchase during a season and shortened distribution channels may compensate the higher price of organic foods. The trend of increased use of ready-to-cook foods does not usually lead to a higher nutritional and health quality, lesser burden on the environment and an economic effect. However, it may be assumed that the expansion of knowledge of catering managers of nutritional quality and environmental impacts, with better experience in optimizing meals and with the proper motivation, parameters of meals in public catering facilities may be combined and thus contribute to the sustainable management in food services.

Keywords: School meals, nutritional quality, environmental aspects, economics, optimization

1. Introduction

The task of school meals is to provide proper nutrition to students during their stay at school and, at the same time, form positive nutritional, hygiene and social habits of students [1]. Generally, school meals should be an example of good nutrition and should make children acquainted with new meals that children do not know from home and, at the same time, teach them the food and dining culture [2]. It also aims at a change of wrong habits that children bring from their families. This includes, for instance, the insufficient consumption of fruit and vegetables, legumes, fish, wrong amount of food, less soup, higher consumption of sweet dishes, dumplings and fatty dishes. Consolidating and acquiring hygiene, cultural and social habits, which include personal hygiene (especially hand washing), the cultural and hygienic rules of dining, a proper use of cutlery, table manners etc., is also a part of this education [3]. Easy accessibility, mostly at the place of school attendance, and subsidized meals, which become available for all social groups, may be included among the positives of school catering. Certainly, mass catering has some disadvantages. These include a limited selection of dishes, poor quality of service (in essence, it is a self-service), often poorer quality of food, smaller portions, eating in haste, also the environment is not usually very calm and the optimal time and duration of a meal are not respected [4]. Catering managers, chefs and service staff, as well as methodological workers and educators, who train personnel for school catering facilities, are in a position to meet the considerable demands made of them due to efforts to eliminate the drawbacks.

The menu is the result of efforts to comply with the set of school food standards and regulations and also an operating plan of the facility for a certain period (usually a month). Menus are drawn up by school catering managers in collaboration with the executive chef in order to suit not only the principles of a healthy diet but also technical possibilities and staff deployment of the facility as well. They should be varied, creative, modern and meet the nutritional recommendations for children [5].

The principle of full use of seasonal market opportunities is very important. An executive chef must be familiar with the offer of foods, especially fruit and vegetables, and their prices. It is also important to take into account the operating conditions of the facility, technical and mechanical equipment of the kitchen, serving system, the number of staff and their qualification, the supply situation when drawing up a menu. The alternation of different cooking techniques is essential as well. Besides meat meals, the meals that contain vegetable protein (soufflés, vegetable, legume and cereal meals), meals accompanied by cheese, cottage cheese, dairy products should be put on the menu. Each lunch should be complemented by a vegetable side dish or a salad (excluding sweet meals), fruit or raw vegetables. In case of a necessary change of the menu, the alternative meal should be similar in the energy and biological content to the originally planned meal [6].

Menus in school catering facilities should be nutritionally balanced, offering tasty and attractive meals to diners not too financially demanding and, last but not least, manageable. The main tasks for the kitchen staff are:

- adhere the energy and biological values of the diet (reducing fat intake or sugar used),

- respect the age categories of children boarding in the facility (nursery, primary, secondary schools)

- take into account the season and the use of seasonal foods,

- provide the diversity of meals in relation to consistency, colour, taste and technological treatment,

- guarantee the greatest possible variety of foods from different groups in order to provide adequate intake of nutrients, vitamins and minerals through: including all kinds of meat – beef, low-fat pork, poultry and especially fish, changing side dishes and different kinds of vegetables regularly and avoiding using the same foods, that undertaken different techno-logical treatment, in one day [7].

2. Literature review

2.1. The nutritional quality of meals

The nutritional quality reflects a content of substances, which has positive effects on human nutrition, their internal composition and proportions. The nutritional role of school lunches involves delivery of about 35% of the recommended daily energy intake [2]. In modern history, there have been changes in eating habits and physical activity. More meat, meat products and sweets are eaten, a lot of sweet, chemically flavoured drinks are drunk, a sedentary lifestyle prevails. Naturally, this lifestyle leads to overweight and obesity. Many school cafeterias and vending machines placed in the corridors of schools, whose range of goods resemble classic fast food restaurants, which children prefer to healthier alternatives offered in school catering facilities, contribute to the unhealthy trend. The main deficiency is the internal structure of meals, often dominated by animal products and an associated excess of animal fat, cholesterol. Another problem is the inadequate intake of certain vitamins and minerals.

The nutritional intervention aimed at the change of technological methods of food preparation, that would still respect traditional Czech cuisine at the same time, appears to be a quick way to make school meals healthier. The intervention program has been running since 1993 and its principles read:

- Meat – use rather less often, but of a high quality and fat-free. Do not use trimmed parts for further processing in school meals, use plant foods (legumes, oat flakes) to get quantity and energy value.

- Milk and dairy products – include as often as possible, choose low-fat products, e.g. in the form of drinks, sprinkles and baking with cheese, salads with yogurt. Provide dairy snacks.

- Vegetables and fruits – with each meal. Prefer raw vegetables (salads, side dishes), favour frozen vegetables to pickles during off-season

- Legumes – generally increase their share in the diet. Offer more frequently and in smaller portions (e.g. adding to soups, minced meat, soufflés and salads).

- Desserts – prefer healthier alternatives based on the processing of dairy products (cottage cheese, custard), use oat flakes, whole meal flour, reduce sugar and fat.

- Fat – keep animal fat to a minimum, use vegetable oils (sunflower), preferably without heat treatment (salads), reduce the use of roux [8].

- The nutritional quality of school meals is based on the recommended nutrient intakes provided in 1989. These focus on the issue of energy demands, the major nutrients and other essential factors for the human body. They are based on the physiological needs of a human body and are calculated for different categories according to age, physical activity and physiological condition [9].

Recommended nutrient intakes are guidelines for creating so-called consumer's basket. It describes the average food consumption calculated from the basic range of foods in the form of "as purchased" (i.e. it takes into account losses, e.g. when trimming vegetables, fruits, etc.). Food consumption is expressed as a percentage and should correspond to the monthly average with allowance of ± 25% [10]. The consumption of meat, fish, milk, dairy products, fruit, vegetables, potatoes, legumes, sugar and fat may be monitored by means of the consumer basket [11]. There is a rule that the average intake of vegetables, fruit, fish and legumes represents the lower limit, which may be exceeded, and the intake of free fat and sugar represents the upper limit, that is desirable to be decreased [10]. Czech School Inspectorate and Regional Hygiene Station monitor if the consumer basket is respected [3]. Recommended nutrient intakes are updated at regular time intervals. Currently, the Czech Republic has adopted a new list of recommended nutrient intakes from the German-speaking Central European countries – the so-called Reference values for nutrient intake (DACH - Deutschland, Austria, Chuisse). These should be taken into account when developing new nutrition standards for school meals. However, setting up new consumer's baskets may not be as fast as it might occur. The reasons are economical, and perhaps political and social as well, also the current eating habits of our population may influence that. The recommended intake of protein is rapidly reduced (from current 2.4 g/kg of body weight to 0.9 g/kg of body weight) according to the DACH; therefore, it may cause some dissatisfaction of the part of diners within our eating habits [9].

The tool to combine different food commodities in order to meet the consumer's baskets is called "the recommended dietary variety". It is not officially set; however, it specifies the number of times in a month a certain type of food should be included on a menu: milk, legumes, fish, etc. [12]. Therefore, not only the fact that the consumer's basket is filled is observed but also the way it is filled in: e.g. preference of lean meat to fatty meat or smoked-meat products, raw or cooked vegetables to pickled, cutting down on sugary and fried meals (max. two per month), the inclusion of sufficient quantity of fish, legumes, substituting conventional side dishes with, for example, millet, buckwheat, couscous, oat flakes, etc., providing fruit and vegetables on a daily basis. The requirement for using different cooking techniques comes from the recommended dietary variety [6]. The different types of dishes should be included usually only once a month. Exceptions are seasonal foods that may be used more frequently. In addition to classic recipes, school catering facilities may use their own or regional recipes but they must comply with all the principles mentioned above [13].

When drawing up the menu, we are limited by the consumer basket and financial limit, and the recommended dietary variety is used as a guideline. Menu is usually drawn up by the catering manager in cooperation with the executive chef for a few weeks in advance, usually for a month, and later it is specified. It must conform not only to the principles of healthy nutrition but also financial, technical and personnel capabilities of the catering facility [5]. If, for any reason, a change is needed, an alternative dish should resemble the originally planned dish in terms of energetical, as well as biological aspects [6].

2.2. Ready-to-cook foods

There have been growing requirements in the area of food preparation, hygiene and final treatment and dining, that modern and classic gastronomy has to meet. There are four basic guidelines to prepare and distribute meals in a public catering facility:

1. Joint catering facility – dishes are prepared in a local kitchen of fresh ingredients, as well as ready-to-cook foods. Capital and operating costs are higher (staff, energy) and facility management must be professionally qualified. The more school uses fresh ingredients, the more hygiene must be respected. Demands on input check of goods, storage and preparation and needs for workspace increase. Preparing meals in their own kitchens is mainly a matter of boarding schools.

2. Cook & Chill - dishes are refrigerated and supplied by a professional food provider or from a central catering facility. Dishes are cooled to 3°C immediately after cooking, may be portioned and then stored in cool conditions (0–3°C) by an external supplier. The staff of the school catering facility provides only heating (which must not exceed 30 min) and distribution. There are strict hygiene requirements for the preparation and storage if the dishes are produced by a central catering facility. It must be cooled within 90 min and should not be stored for more than 3–5 days until being re-heated. Cooled products are used mainly at secondary schools.

3. Frozen system – a professional provider provides frozen dishes as individual portions or the whole menu. They are frozen to –18°C after cooking and the temperature is maintained during the transport. The cooling chain from the producer to the final treatment before being served must not be interrupted. Workers of a school catering facility provide heating; meals may be portioned for serving where necessary. After that meals must be continuously served. The advantage of this system is that the necessary investments to draw up a menu are low, as well as low demands on the qualification of staff. The system is mainly used at secondary schools.

4. Hot meals – dishes are provided already completely ready and warm by an external supplier – a catering company that provides distribution in the facility as well. Each serving of dish is put into a thermo box or a food container (larger amount) immediately after cooking. Thermo boxes retain the internal temperature of 70°C from the filling to the

distribution of meals. The temperature when served is then 65°C. For cold foods, the temperature should be in the range of 8–10°C. The system of hot meals (60%) dominates, followed by the joint catering facility (about 20%) at the full-time German schools.

In the Czech Republic, the system of the joint catering facilities still clearly dominates. In Europe and around the world, there are significant differences in terms of the range of school meals, support and forms of preparation and distribution of meals. The differences result from the traditions, economic strength and social policies of individual countries. Globalization trends have brought an increase in the use of ready-to-cook foods, convenience foods and ready-to-eat meals, which always have a higher degree of processing than the base material, in a number of countries and in the Czech Republic as well. These dishes or foods, convenient for immediate consumption, are in most cases frozen, canned or dehydrated and therefore they must be somehow processed before consumption. The importance of using ready-to-cook foods has its benefits, especially in terms of time savings needed to prepare, workforce and costs, they extend the range of dishes, which would be difficult to prepare in ordinary kitchens, support the creativity of a chef. Some facilities are unreasonably mistrustful of these foods and products. Partially, they may be put off a higher price of the ready-to-cook foods, even though the difference is relative in many cases. It is worth being aware, however, when the use of ready-to-cook foods is appropriate and in what cases we may do without them. Chlumská [14] points out a finding that the use of ready-to-cook foods or ready-to-eat meals is one of the most common complaints against the school board from conscious parents. According to her, school catering facilities tend to use these products partly because the market offers an increasingly greater choice, as well as due to reduction of staff, when school catering facilities must provide the preparation of meals with fewer employees than before.

2.3. Economic aspects of school meals

Depending on how the school board is managed and how the state and municipalities participate, European countries may be divided approximately into three groups: the first one includes the states where school meals are provided to children for free (Finland and Sweden), the second one includes the states where school meals are organized centrally or regionally in some way and the costs are partly covered by the state or municipalities (France, Belgium), and finally in the third group there are states where school meals are not uniformly organized or not implemented in the way that we know in the Czech Republic [15].

School meals are not based on profit in most countries, thus differ from conventional manu-facturing company in a market economy. Therefore, costs are one of the most important criteria and affect pricing greatly. The cost of providing food service may be, in terms of the types of costs, divided into the costs of foods, personnel costs – salaries, training and social statutory costs and operating costs – energy, other materials, services, depreciation, etc. [16].

In countries that support school meals, diners only cover a portion of the actual price of the meals. School facilities in the Czech Republic must follow nutritional standards, the average consumption of foods and financial specifications for the purchase of foods for each age group. The part of the price of a meal paid by parents (i.e. the price of foods) may be set differently based on an agreement with parents at private schools. At schools that are run by municipalities, county or state, the price of foods is limited by so-called financial specifications, which are specified in the school food regulation [14]. The set financial specification must amount to the sum that enables a school catering facility to meet the requirements for the consumer's basket. It also specifies the financial spread – an amount of money that school facilities may use to make a lunch – i.e. soup, main course, salad, dessert and beverage [15]. At present, the cost for foods to make a lunch for one diner ranges from 14 to 37 CZK, which corresponds to 0.5 – 1.2 Euro, in the Czech Republic.

Personnel costs include wages and salaries of the employees of the facility, their further education and working instruments and are funded from the state budget.

Energy consumption, costs of services, costs of other materials and depreciation of tangible assets make up a significant portion of operating costs. These costs are covered by the institutor. Although the amount of personnel and other operating costs are based on a calculation, it is not a normative expense but a cost that may be influenced by an effective and efficient use of available resources [16].

From an economic point of view, the quality of school meals may be influenced in a few ways only, virtually through bargains, donations or grants as extra sources of money [1]. The more diners of a facility, the easier it may be to achieve beneficial agreements or quantity discounts for ordered foods. Purchase of seasonal foods, especially fruits and vegetables, is another way to influence the price of foods and respect the nutritional standards at the same time. Their price change regularly according to a season and thus to their availability. The money saved on purchase may be used to enrich and improve (pot. make cheaper) the diet [17].

2.4. Environmental aspects of school meals

Our eating habits are created especially in the context of public catering. High-quality and healthy foods in catering facilities show not only the value chain of diners but also an environmental responsibility. A sustainable economic system must support especially environmental-friendly regional production and consumption of fresh natural foods.

Food production uses an increasing amount of energy with a corresponding negative impact on the environment. An important factor is the origin of foods, resp. transport distance from a producer to a consumer. A reduction in the proportion of meat on the menus and consumption of regional vegetable products allows caterers to reduce the impact on the environment. The negative impacts of the use of ready-to-cook foods or ready-to-eat meals, processed products and products stored for a long time outweigh their benefits due to the heating and cooling of foods, special packaging and transport costs [18].

Research shows that the use of local, seasonal and organic foods and the preparation of fresh meals of them may significantly reduce the proportion of greenhouse gas emissions (GHG) in catering facilities.

An indirect energy consumption, i.e. energy that comes from foods, their production, processing and trade, constitutes up to 63% of total GHG emissions in catering facilities. The largest amount of GHG comes from meat in catering facilities. The use of meat and meat products in Austrian catering facilities makes up 14% of the total amount of the foods, therefore 63% share of GHG emissions in the indirect energy consumption is very high.

The implementation of sustainable diets and thus optimized meat portions and increases of the share of vegetarian dishes have also saving potentials within GHG emissions. Vegetarian dishes produce up to 99% less GHG emissions in comparison with meat dishes. Also the use of regional and seasonal foods and organic foods makes emissions savings. Local foods have the potential to save up to 50%. Using foods from an organic production can reach up to 40% savings. A level of food processing plays an important role in addition to the criteria of regionality, seasonality and organic farming with regard to the GHG emission topics. Each step represents a further production of GHG. One kilogram of fresh conventional potatoes produces 0.31 kg CO^2eq, but one kilogram of potato chips produces 4.36 kg of CO^2. The trend of an increased use of ready-to-cook foods in catering facilities has primarily economic reasons (e.g. less staff needed). However, this is often compensated by a greater need for goods. Constant heating and cooling, special packaging and food miles (mileage when transporting food to the kitchen) and often questionable additives as well have negative effects on the environment [18].

2.5. Local foods

School catering facilities are one of the major purchasers of local products [19]. The reason for the preference of local foods is that these foods are much fresher due to short distribution routes than the foods that take long-distance routes. Therefore, fresher local foods generally tend to taste better and more valuable nutritional parameters. The fact that the closer the food is to the consumer, the lesser burden on the environment during their transport is also important. Reduction in the proportion of meat on the menus and consumption of local vegetable products allow caterers to significantly reduce environmental impacts, as well as take into account the financial aspect (Eagri-Regionální potravina, 2009–2013). An extension of the path that an agricultural product takes from the producer to the consumer may lead to a loss of authenticity. Consumers and also control bodies may supervise the foods produced in local conditions better and thus there is an indirect pressure on producers to maintain the quality of their products at a high level. Another reason for the preference of local foods is that these foods are much fresher due to short distribution routes than the foods that take long-distance routes. Therefore, fresher local foods generally tend to taste better and more valuable nutritional parameters. The fact that the closer the food is to the consumer, the lesser burden on the environment during their transport is also important [20]. A significant aspect to prioritize local foods is that it promotes employment in the region. Then prosperous farmers, processors and vendors represent a guarantee of maintaining or even expanding the number of jobs.

2.6. Organic foods

Reasons for the introduction of organic foods in schools are mainly attempts to encourage children to eat healthier and better diet. Equally important is the positive impact on dietary habits and a healthy lifestyle. Organic foods are not used in school catering facilities in the Czech Republic very frequently. Currently, it is estimated that approximately 150–300 kindergartens and schools use organic foods in significant quantities, which represents about 1.5–3% of the total 10,500 schools (nursery, primary and secondary schools). The schools that have participated in one of the pilot projects for the introduction of organic foods in schools or alternative schools (especially Waldorf kindergartens and schools), where the use of organic food is a part of their philosophy, have been ahead [14]. The reason for the low interest in organic foods is their high price. Currently, no financial subsidies for their purchase are provided [21]. However, the price of school meals in the school catering facilities, which have introduced organic foods, has increased only very moderately by about 10%. Organic cereal products, legumes and dairy products are used most often. Conversely, baked goods, meat and meat products and other products are used in the smallest amounts in schools. Many countries have supported the use of organic foods in schools and other public catering facilities in various ways including legislative measures, subsidies and other incentives. For example, the Italian government has adopted a law requiring the use of organic products in school catering facilities. Therefore, the Italian legal system" creates direct and explicit relationship between local organic products and catering services." This national law has created an environment that encourages many municipal authorities to start purchasing organic products. The support of catering facilities, while optimizing diets that account of local, seasonal, fresh and organic foods, will enhance regional economic structures, potential energy savings in catering facilities and offer healthier boarding

3. Objective of the study

The main objective of the UMBESA project is to support catering facilities when introducing sustainable diets. This can be achieved by increased use of organic, local, seasonal and fresh foods and reducing meat portions. These steps should support not only the environmental protection but also physiological and optimal nutrition. The project consisted of five main parts. The first part focused on the current consumption of foods and diet composition in school catering facilities, these documents should establish a basis for change. The second part dealt with the evaluation of similar projects, which aimed to introduce the above mentioned criteria towards sustainable diets and the objective was to identify the strengths and weaknesses of these projects. The third part aimed to identify the current networks of suppliers of school catering facilities and stakeholders who are involved in the field of public catering, at the same time, new stakeholders were identified and a new network, meeting the sustainability criteria (e.g. regional and organic suppliers), was proposed. The fourth part of the project had as its object assessment of opinions on the current state of catering services and the state after introducing some changes (see the fifth part of the project), a survey had been carried out. The fifth part of the project dealt with the actual implementation of changes and it is discussed in this chapter as the main output of the project.

As described before, the aim of the experiments within the project is an active support of the introduction of sustainable diet in catering facilities. In selected school catering facilities, certain recipes were chosen (hereinafter original dishes) and modified (hereinafter optimized dishes) according to the criteria of sustainability (an introduction of ecological, local and fresh foods and a reduction in meat portions). These two dishes were evaluated and compared within the selected criteria. The aim of this part was to assess whether a change of diet contributes to sustainability and also answer the following questions:

- What measures can be realized in catering facilities to optimize towards sustainability?

- What economic, ecological and nutritional–physiological positives and negatives arise in catering facilities using sustainable foods?

4. Methodology

Methodical procedure briefly describes the methodology of the individual parts of the project, with the greatest focus on the methodology of experimental cooking and their evaluation.

4.1. Analysis of foods and menus

Lists had been drawn for each school, which grouped foods into appropriate groups using the annual statement of the shopping list of foods for the reference year of 2011, which included the price of foods, as well as their suppliers. At the same time, the lists had been drawn up and assessed according to their origin – regionality of foods, their seasonality, processing – frozen, fresh and ready-to-cook foods and also from the perspective of organic production. Furthermore, the lists of dishes according to the proportion of main ingredients – meat, vegetarian and sweet, as well as proportions of organic ingredients, ready-to-cook foods and local ingredients, had been drawn up according to the menu.

4.2. Search of similar projects

Two Austrian, two Czech and two international projects were selected to map out the initial conditions, implementation and factors for success and failure. The authorized representatives of these projects were interviewed; the interviews were subsequently evaluated and reduced in accordance with the summarizing criteria. The analysis according to Kotter's 8-Step Change Model "Leading Change" [22] was performed. The supporting factors, as well as inhibiting factors of success, were found.

4.3. Networking

In the first instance, the current network of suppliers in various catering facilities was identified as a part of search of the ingredient consumption, see Section 4.1. As a second step, a potential supplier network was found and an extensive list of suppliers in various regions and districts was drawn up. At the same time, the selected suppliers were questioned regarding their attitudes to the issue of regionality and seasonality of offered products while creating the potential network. The last activity within networking was to create groupings of regional participants and set up the Steering Committee of the project that discussed the progress of the project and inclusion of dissemination of the results of individual project activities at regular meetings.

4.4. Survey among diners

The survey was carried out in the form of two questionnaires, one at the beginning and another one at the end of the project. The questionnaire included topics such as satisfaction with the catering facility, with its offer, attitude of staff, questions about eating habits of the respondents and, in conclusion, inquiries concerning the project itself. Descriptive statistics, factor and group analysis had been used to evaluate the results and a profile of borders that may be used to propose specific changes to catering facilities was set.

4.5. Experimental cooking

In the fifth part of the project, practical experiments in the context of experimental cooking were carried out, where an original and optimized dish was cooked and mutually compared. The recipes for the original and optimized dishes were presented and recorded by the chosen catering facilities. Relevant data including the preparation of foods, recipes, cooking process (time, equipment used, number of employees, water consumption) were collected during each cooking. The dishes were evaluated from several different vantage points.

4.5.1. Environmental assessments

Ecological assessment was performed by analyzing CO^2 emissions. CO^2 emissions of foods that had been identified within the SUKI project [23] were used as baseline data. The emission burden data of foods that had not been investigated within the SUKI project were complemented by the literature and the GEMIS database search. CO^2 emissions were determined within the ingredients that are most important in terms of quantity. It was necessary to determine CO^2 emissions by at least 50% of the ingredients for one dish.

4.5.2. Economic assessments

Economic assessment was performed by analyzing costs. The following costs were included into the analysis:

• Cost of foods: the current prices of foods including VAT were taken into account.

• Personnel costs: the period of active work was multiplied by the average hourly wage and the number of persons.

• Operating costs: i.e. costs of water and energy.

4.5.3. Nutritional–physiological assessment

The calculation of nutrients was made with the help of a nutrition consultant. The production method (biological, conventional) was not taken into account within the nutritional–physiological assessment. Original and optimized dishes were compared with respect to the amount of calories, protein, fat, carbohydrates and fibre.

4.5.4. Organic – Regional – Seasonal

The proportions of biological, local and seasonal ingredients were determined within the original and optimized dishes.

4.5.5. Qualitative assessment

A sensory evaluation test was used. The test includes food tasting carried out either by the staff themselves or by diners. The results were discussed with the managers of the catering facilities.

5. Results

This part briefly describes the main results of each stage of the project with the greatest focus on the assessment of the experimental part of the project, i.e. experimental cooking.

5.1. Analysis of foods and diets

- The analysis of food consumption in Czech catering facilities showed that the most used group of foods is vegetables (including potatoes) at 34%. The other most commonly used group consists of the cereal products at 16%. They are followed by meat and meat products, as well as dairy products at 14%. The proportion of fruit is 11%. The last group at 12% includes other products.

- The proportion of fresh ingredients is on average 78%, 6% of frozen ingredients and 16% of ready-to-cook products.

- Currently, organic foods are not used in Czech catering facilities or they are used in quantities of less than 1%. That is due to a limited budget for foods and prohibitive costs of organic foods. This corresponds to the total organic food market situation in the Czech Republic, which has not been sufficiently developed yet, the share of organic production on arable land is still too small to successfully compete with conventional products in catering.

- The proportional share of seasonal fruit and vegetables varies from 30 to 90%. It reaches 47% on an average. Undoubtedly, the potato consumption is the biggest item accounting for about 60%. Another important item consists of onions, cabbage, carrots, tomatoes and cucumbers. From fruits, the most important are apples and plums of our domestic production. It is worth noting that the second most frequently used fruits are bananas, which do not meet the criteria of sustainability, both seasonal and local, and it would be good to substitute them with domestic fruit.

- The proportional share of local products varies from 17 to 86%. The average is 39%. The analysis shows that the catering facilities in bigger cities use less local products than the catering facilities in smaller towns, logically, the reason for that is a larger food market and offer in bigger cities. From the local production, meat, dairy products, cereal products, fruit and vegetables prevail. Most ready-to-cook and frozen products have their origins outside the region.

- The analysis of main meals shows that 62% of the main meals are meat meals. Vegetarian meals make up 21%, fish meals 7% and sweet meals 9%.

5.2. Search of similar projects

Based on a detailed analysis of six successful projects (Kuratorium of Vienna Retirement homes, the project in the catering facility of the Lower Austrian provincial office, German restaurant ESPRIT, Italian project iPOPY and the Prober Union, two Czech projects "Organic food for schools" and "School full of health"), there are these fundamental factors of success:

- Use of external influences for change (e.g. childhood obesity).
- Explanation of the meaning of the project to stakeholders.
- Extensive information campaign.
- Setting realistic and achievable targets in the short term.
- Perseverance despite the initial failure.
- Setting goals for the future.
- Building long-term relationships between the entities.
- Adapting the project to existing habits and structure.
- Constant communication with stakeholders.
- Gradual implementation of measures, smooth implementation of the objectives.
- Gaining supporters during the project.
- Value conviction of a person in chargé.

5.3. Networking

An important outcome of networking was a catalogue of ingredient suppliers in each region that was provided to catering facilities in order to enable them to obtain ingredients from local suppliers. The project had been also promoted and consulted within the Steering Committee composed of representatives of the government, experts and business leaders. The survey among suppliers resulted in the following main conclusions:

- It is very important to document the origin of products according to the surveyed suppliers. About 74% of interviewed producers expect that the regionality becomes a sales argument in the future. The amount depends primarily on the size and trade tendency of the producer. Smaller producers try to show the quality of their products using the regionality. The current problem is too many regional brands, which people may find confusing, as well as selling products under a foreign brand and a lack of awareness about the quality of local foods. More than a half of respondents think that the regional origin does not affect the price.

- The seasonality issue concerns mainly fruit and vegetable producers. A large group of the interviewed producers rely on stable buyers who are familiar with seasons when different kinds of fruits and vegetables ripen; therefore, they do not need to be further informed. They do not intend to include the seasonality as the sales argument.

- Regarding the expansion of product diversity, 70% of interviewed producers draw up their offer not concerning reactions of consumers. If we evaluate the cooperation of the producers, we find out that most of them have both stable and vague relationships, as well as regionally focused relationships, because these groups complement each other and eventually intersect, for example, when a customer becomes a stable client.

- It is gratifying to note that most local producers have an increasing interest in their products and that the society slowly begins to realize the true quality and value of local products.

5.4. Survey among diners

About 703 diners of participating Czech catering facilities participated in the first wave of the survey and 713 diners in the second wave. Overall, it may be summarized that their satisfaction with the catering facility, its atmosphere and quality of food had increased.

5.5. Experimental cooking

At least three experimental cooking of original and optimized meals, which were compared using several criteria, took place in each partner catering facility. As an example, the experimental cooking of tomato sauce with beef is being described here.

The original meal consisted of classic tomato sauce with beef and bread dumplings. The optimized meal included a reduced portion of meat and turkey meat substituted for beef, couscous for bread dumplings and some of the ingredients in an optimized meal came from organic production.

5.5.1. Economic assessment

The analysis shows that the costs of optimized meal are by 17% higher. The price per serving is 0.2 EUR higher. More expensive are especially the costs of ingredients and personnel costs, it is due to a greater need for active involvement of staff. Conversely, operating costs are lower because simpler technological demands for preparation dominate.

5.5.2. Environmental assessment

Ecological assessment shows about 69% smaller environmental burden when cooked optimized meals. The ingredients for the original and optimized meal in the total proportion of 99% were included into the assessment.

5.5.3. Nutritional–physiological assessment

One portion of the original dish contains 601 calories, 33 grams of protein, 11 grams of fat, 96 grams of carbohydrates and 4 grams of fibre. A portion of the optimized dish contains 513 calories, 35 grams of protein, 14 grams of fat, 63 grams of carbohydrates and 5 grams of fibre. The nutritional values were taken from nutritional tables. The percentage difference of indicators is shown in Figure 1.

Nutritional-physiological assessment of "tomato sauce"

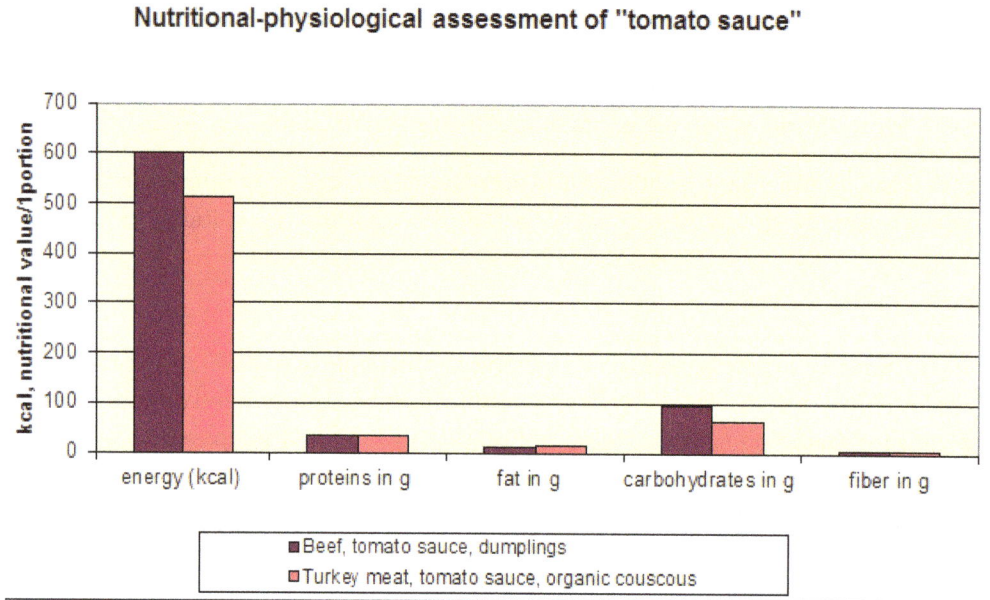

Figure 1. Nutritional-physiological assessment of tomato sauce

5.5.4. Assessment according to the production method (organic, conventional)

100% of ingredients for the original dish were produced in conventional agriculture, whereas the proportional share of organic ingredients in the optimized meal is 23%.

5.5.5. Assessment according to the processing method (fresh, frozen or ready-to-cook)

Both the original and optimized dishes do not contain frozen ingredients and consist of fresh and ready-to-cook ingredients only. The proportional share of fresh ingredients is 44% in the original dish and 86% in the optimized dish.

5.5.6. Assessment of seasonality

Seasonality is assessed for vegetables and fruit, the original dish contains onion and the optimized dish contains onion and tomatoes. The original dish may be described as seasonal in the months of May, June, July, August and September. The optimized dish may be described as seasonal in the months of June, July, August, September and October.

5.5.7. Assessment of regionality

To assess the regionality, the origin of main ingredients of a meal was determined as a percentage, i.e. that the percentage of these ingredients constituted at least 80% of the meal. Regionality of ingredients may vary during the year, depending mainly on a purchase of seasonal ingredients. The original dish contains almost no seasonal products and the suppliers remain the same throughout the year and the proportional share of local ingredients is 35%. The optimized dish contains 37% of local ingredients in the months from June to September, whereas in other months it is 0%.

5.5.8. Qualitative assessment

Ten employees of catering facility answered in the carried survey that the original dish leads in the overall ranking, but also scores in the individual categories of taste, smell and appearance better than the ready-to-eat meal. The results are shown in Figure 2.

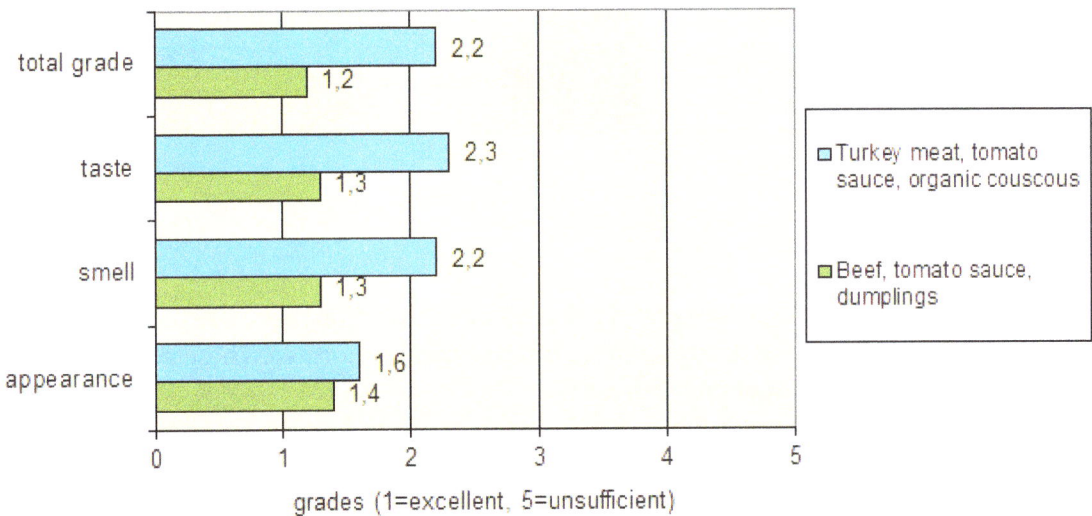

Qualitative assessment of "tomato sauce"

Legend: Turkey meat, tomato sauce, organic couscous; Beef, tomato sauce, dumplings

total grade: 2,2 / 1,2
taste: 2,3 / 1,3
smell: 2,2 / 1,3
appearance: 1,6 / 1,4

grades (1=excellent, 5=unsufficient)

Figure 2. Qualitative assessment of tomato sauce

5.5.9. Results of other selected dishes

Table 1 shows the results of other selected experimental cooking. The results in each column are always related to the optimized meal. The costs column shows the difference between costs of the optimized meals per serving, the CO_2eq column evaluates the environmental burden, i.e. the difference in the amount of produced greenhouse gases and the share-of-organic-ingredients column and the share-of-fresh-ingredients column display the difference in proportion of organic and fresh ingredients. For the sake of clarity, the aspects identifying areas of improvement are marked in green, aspects that show deterioration are marked in red and aspects with no indication of a change are yellow.

Original meal	Optimized meal	Costs	CO$_2$eq	BIO-share	Share of fresh foods
Risotto with vegetables and pork	Couscous risotto with vegetables and chicken	+ 24%	+ 2%	+ 31%	–2%
Pork goulash with dumplings	Bean goulash with bread rolls	–3%	–41%	-	–66%
Fillet with potatoes	Carp with potatoes	+ 45%	–21%	+ 13%	+ 15%
Meat rolls with mashed potatoes	Meat rolls with spinach and tricolour rice	–7%	–35%	-	–13%
Meatball with mashed potatoes	Burger with broccoli and cheese and mashed potatoes	+ 24%	+ 47%	-	+ 2%
Stuffed cabbage leaf, potatoes	Cabbage leaves stuffed with buckwheat, potatoes	–7%	–18%	-	+ 180%
Fried meatballs	Buckwheat burgers	+ 16%	–65%	-	–26%
Bread pudding with cream cheese	Bulgur with fruit and raisins	+ 12%	–74%	-	–51 %

Table 1. Results of experimental cooking

5.5.10. Discussion on meal optimization

There were a total of 32 experimental meal preparations, whose aim was to compare the original and optimized meals in several respects, had been performed. These general conclusions result from the assessment of each meals:

- Economic perspective: It always depends to what extent the original meal was modified, e.g. costs may be reduced when meat portion sizes reduced significantly, on the contrary, the increase of costs may be connected with the use of organic foods and some fresh and local foods (e.g. using fresh carp instead of frozen cod), the highest price increase was in our case by about 45%, the highest price reduction was by about 78%, the optimized meals are on average by 2% more expensive. Some conducted studies (e.g. results of the project "Organic food for schools") show that consumers have an interest to pay more for quality.

- Environmental perspective: Most of modified meals result in decreased production of greenhouse gases and thus a positive environmental effect. The most significant reduction was by 88%, the greatest increase was by about 345%; however, this figure is completely beyond the average increase in emissions, which makes up approximately 20%. Putting this excessive result aside, emissions of the optimized meals decreased by 74% on an average.

- The proportion of organic ingredients: Regarding the share of organic foods in recipes, only a small proportion of experimental cooking included such foods. In particular, dry foods, alternative foods such as bulgur, then vegetables and in one case meat were used. However, the inclusion of organic foods, particularly meat, meant an increase in the price of meal. This fact is due to the current state of the organic food market, where their prices are still significantly higher than the prices of their conventional analogies.

- The proportion of fresh ingredients: The proportion of fresh foods had increased significantly at the expense of the ready-to-cook foods. The average increase reached 90%.

6. Conclusion

The diet structure of monitored school catering facilities shows that the normative indicator of the nutritional quality of food (consumer basket) is respected. Traditional meat dishes (62%) prevail, the trend of vegetarian diet is slow, but positive. Seemingly satisfactory representation of vegetables in a diet is given by traditionally high consumption of potatoes. Organic foods are almost absent in Czech school catering facilities. The reason is high price and low availability. Great variability in the consumption of local foods (17–86%) and seasonal foods (30–90%) indicates significant reserves for suppliers and catering facilities. Larger facilities tend to use ready-to-cook foods and ready-to-eat meals more. Their origin is mainly supraregional. Greater use of local, organic, seasonal and fresh foods is possible, thanks to the relationship between producers and food distributors in the region. Optimizing rarely improve all the required parameters, particularly difficult is to coordinate economic aspects with an ecological criterion, as well as use of fresh, local and organic foods. However, in many cases, improvements in the above mentioned parameters did not mean a noticeable increase in prices. Motivation of staff and consumers towards sustainable diet is a long process that requires continuous awareness of both parties. School meal plays an irreplaceable role in education in a healthy lifestyle.

Acknowledgements

Supported by the University of South Bohemia in České Budějovice grant GAJU 063/2013/Z.

Author details

Jan Moudry Jr, Jan Moudry[*] and Zuzana Jelinkova

*Address all correspondence to: jmoudry@zf.jcu.cz

University of South Bohemia in Ceske Budejovice, Czech Republic

References

[1] Indrová J, Petrů Z, Vaško P. Podnikatelská činnost ve stravování a hotelnictví. 1st ed. Praha: VŠE; 1995. 108 p. ISBN: 80-7079-054-7

[2] Laláková H. Stravovací zvyklosti dětí na základní škole. Absolventská práce. Praha; 2000. 49 p.

[3] Kotulán J, et al. Zdravotní nauky pro pedagogy. Brno: MU; 2005. 258 p. ISBN: 80-210-3844-6.

[4] Pokorný J, Pánek J. Základy výživy a výživová politika. 1st ed. Praha: VŠCHT; 1996. 138 p.

[5] Šulcová E. O jídelníčku. Výživa a potraviny: Zpravodaj pro školní stravování. 1999;2:20–20. ISSN 1211-846x

[6] Hnátek J. Výživa a stravování žáků základních a středních škol. 1st ed. Praha: SPN; 1992. 320 p. ISBN: 80-04-23948-x

[7] Packová A. Nutriční a gastronomické zásady sestavování jídelních lístků. Výživa a potraviny: Zpravodaj pro školní stravování. 2010;2:20–20. ISSN 1211-846X

[8] Hrubý J. Jaký je vlastně přínos školního stravování pro výživu dětí a mládeže. Výživa a potraviny: Zpravodaj pro školní stravování. 2001;2:18–18.

[9] Jídelny.cz. Informační portál hromadného stravování [Internet]. 2015. Available from: http://www.jidelny.cz/show.aspx?id=1534 [Accessed: 2015-04-15]

[10] Adamec V, Bártlová E., Černá M. Životní podmínky a jejich vliv na zdraví obyvatel Jihomoravského kraje. Brno: Grada; 2006. 45 p. ISBN 80-7066-969-1

[11] Slavíková M, Vlčková L, Skorkovský J. Screening nutriční úrovně školního stravování v české republice [Internet]. 2010. Available from: http://apps.szu.cz/svi/hygiena/archiv/full10/h2010-3-02-full.pdf [Accessed: 2015-04-15]

[12] Štajnochrová S; Petrová J. Doporučená pestrost v jídelním lístku. Jídelny.cz. Informační portál hromadného stravování [Internet]. 2011. Available from: http://www.jidelny.cz/show.aspx?id=1105 [Accessed: 2015-04-15]

[13] Výživadětí.cz [Internet]. 2009. Available from: http://www.vyzivadeti.cz/pro-rodice/hodi-se-vedet/cim-se-musiskolni-jidelny-ridit.html [Accessed: 2015-06-02]

[14] Chlumská L. Biopotraviny ve školních stravovacích zařízeních – analýza [Internet]. 2009. Available from: http://eagri.cz/public/web/file/39454/brozura_Bioskoly_11_2009_CD.pdf [Accessed: 2015-06-02]

[15] Věříšková L. Školní stravování u nás a v zahraničí. Výživa a potraviny: Zpravodaj pro školní stravování. 2005;2:162–162.

[16] Furstová A. Legislativa školní jídelny. Výživa a potraviny: Zpravodaj pro školní stravování. 2013;4:53–55.

[17] Johanidesová O. Co děti rády jedí ve školní jídelně. Výživa a potraviny: Zpravodaj pro školní stravování. 2006;2:29–30. ISSN: 1211-846x

[18] Daxbeck H, et.al. Nachhaltiger Speiseplan–Umsetzung der Nachhaltigkeit in Großküchen unter besonderer Berücksichtigung von regionalen, saisonalen, biologischen Lebensmitteln und frisch zubereiteten Speisen, zusammenfassung der wichtigsten ergebnisse aus Österreich und Tschechien. Wien: RMA; 2014. 73 p.

[19] Stará A. Regionální jídla ve školní jídelně. Jídelny.cz: Informační portál hromadného stravování [Internet]. 2013. Available from: http://www.jidelny.cz/show.aspx?id=1383 [Accessed: 2015-06-02]

[20] Moudrý J, et al. Realizace trvalé udržitelnosti ve stravovacích zařízeních se zvláštním zohledněním regionálních, sezónních, biologických potravin a čerstvě připravovaných jídel: závěrečná zpráva projektu UMBESA. České Budějovice: ZF JU; 2014. 55 p.

[21] Ludvík P. Školní jídelny v ČR a jejich obraz v médiích [Internet]. 2010. Available from: http://www.jedalne.sk/sk/public/obraz_v_mediich.pdf [Accessed: 2015-06-02]

[22] Kotter, J.P. Leading Change. Boston: Harvard Business School Press; 1996. 187 p. DOI: 10.1002/cir.3880080221

[23] Daxbeck H., et al. Možnosti stravovacích zařízení k redukci emisí oxidu uhličitého (Projekt SUKI). České Budějovice: ZF JU; 2011. 59 p.

5

Pollution Prevention, Best Management Practices, and Conservation

Maliha Sarfraz, Mushtaq Ahmad,

Wan Syaidatul Aqma Wan Mohd Noor and Muhammad Aqeel Ashraf

Additional information is available at the end of the chapter

Abstract

Farming imposes unenthusiastic externalities upon society. It effects by different sources such as loss of biodiversity, land erosion, nutrient overflow, more water usage and pesticides. Optimistic externalities include respect of nature, independence, free enterprise, and the quality of air. Natural methods decrease some of these costs. It has been proposed that organic farming can reduce the level of some negative externalities from (conventional) farming. Organic farming seems to be more appropriate as it considers important aspects such as sustainable natural resources and the environment. For sustainable agriculture, the most important key is the conservation of natural resources. As natural resources become increasingly short in supply, in the coming years the transition to a more resource-efficient economy must be a top priority. Agriculture is the most important sector for ensuring food security for next generations while decreasing the resource use and increasing resource recycling. Various studies have been conducted to compare organic and conventional farming systems and the result shows that organic techniques are less damaging than conventional ones because of the decreased level of biodiversity, less use of energy, and lesser amount of waste production. The researchers of various studies concluded that comparing conventional and organic farming demonstrated that organic agriculture poses lower environmental impacts. However, researchers believe that the perfect result would be the expansion of ways to produce the uppermost yields possible by the combination of these two farming systems and to develop the new system for environment, land, and sustainable forests. Biodiversity from organic farming provides assets to humans. Species found in organic farms increase sustainability by decreasing human inputs such as pesticides and fertilizers.

Keywords: Organic foods, externalizes, environment, impact assessment, conservation

1. Introduction

At present, across the world, industrialized and industrializing countries are consuming the earth's resources at an alarming rate. The world population is continually on the climb. More people on earth and changing their consumption pattern increase their essential requirements for more basic human needs like food, water, shelter and energy. This leads to suggest that an essential rethink of the way we manage our natural resources.

Rising means of agriculture farming is the reason that human lives in the world today. For survival these are the necessary means without which there would be famines all over the world. From many thousands of years agricultural farming was a natural process that did not harm the land it was done on. Farmers used such methods for agriculture that after passing of many generations soil would still be fertile as ever, while modern agricultural practices have started the process of agricultural pollution and this causes the degradation of land, environment and ecosystem due to by-products of agriculture. No particular cause can be credited to the extensive agricultural pollution we face today. Agriculture is a multifaceted activity in which the growth of crops and livestock has to be balanced completely. Agricultural pollution progression stems from the many stages their growth goes through.

To be well thought-out a best management practice, an action is required which increase the crop production while reducing the impact on environment. This means that for healthy crop using the best management like reducing the pesticide treatment. Soil plays a very important role for healthy crops and its management is very necessary, it may be challenged by intensive production of horticultural crops. Farming technologies degrade the natural resource base because they require high toxic chemicals. Organic farming rely on the management of soil organic matter to increase the physical, biological and chemical properties of soil for optimization of crop production. Soil management controls the supply of nutrients to the crops. Soil processes furthermore play a key role in suppressing the pests, weeds and diseases. Agricultural research based on technology should be developed by specialist and then transferred to the farmers through demonstration. Environmentally friendly farming system relies on minimal chemical use like pesticide and herbicide because they play an important role in erosion control. Several authors have already described the potential effects of conventional farming versus organic farming on soil erosion control (Lotter etal, 2003; Erhart and Hartl, 2010, Goh, 2011).

The International Federation of Organic Agriculture Movements standards suggested that by using the minimal tillage, crop selection criteria, maintenance of soil plant cover and other methods which reduce the soil erosion, organic farmers should reduce the loss of top soil cover for better production of crops. Conservation tillage should be adopted by organic farmers especially if they are located in areas susceptible to erosion (IFOAM, 2000). The nutrient contribution is very important in organic farming. By organic manure and rotation nitrogen fixed in the legumes and supplied to the crop. Tillage is also very important because it contributes in incorporation and distribution of nitrogen in the topsoil (Koepke, 2003). This chapter explores how organic farmers can utilize a range of management practices to develop and maintain the soil fertility in order to achieve these wider goals.

1.1. Organic farming

In organic farming, food is grown and processed using no synthetic fertilizers, but pesticides derived from natural sources may be used in producing organically grown food (NOSB 1995). Organic farms reduce some of the negative impacts of conventional farming such as soil erosion and leaching of carbon and nitrogen [1-3]. Organic production has been practiced in the United States since the late 1940s. From that time, the industry has grown from experimental garden plots to large farms where products are formed and sold with specific organic labels. More than forty different state agencies currently certify organic food but their standards are different. According to the organic food production act of 1990, there would be a national list in which the synthetic and non-synthetic substances mentioned cannot be used in organic farming. Organic farming can contribute to protect the environment and nature conservation [4-5].

1.2. Principles of organic agricultural

- Organic farming or agriculture contributes to the health and well-being of plants, animals, soil, earth, and humans; it also provides the nourishment of ecological, physical, and social welfare as it provides chemical- and pollution-free food for humans.

- Equality is obvious in maintaining the integrity of the joint planet mutually amongst humans and further living beings. It is helpful in decreasing poverty and improves the quality of life.

- In the living ecological system, organic farming must be modeled because these methods fit the environmental cycles and equilibrium of the natural world.

- Natural farming should be accomplished in a vigilant and accountable way to promote the environment and generation at present and in the future.

1.3. Regulations for organic farming

The National Organic Program proposed some regulations that will ensure that organically labeled products meet consistent national standards.

- Any farm crop harvesting or handling operation that wants to sell an agricultural product as organically produced must adhere to the national organic standards.

- The national organic standards for production process address the methods, practices, and substances used in producing and handling crops, livestock, and processed agricultural products.

- Organically produced food cannot be produced using excluded methods, sewage sludge, or ionizing radiation.

- The organic crop production standards say that land will have no prohibited substance for 3 years before organic crop harvesting, no use of genetic engineering and ionizing radiation, soil fertility and crop nutrients will be managed, organic seeds and planting stock will be preferred, crop disease, pests, and weeds will be controlled.

- In the livestock standards, slaughtering of animals must be raised under organic management, organically raised animals may not be given hormones to promote growth, and all organically raised animals must have access to the outdoors, including access to pasture for ruminants.

- The handling standards say that all non-agricultural ingredients must be included on the National List of Allowed Synthetic and Prohibited Non-synthetic Substances.

1.4. Environmental benefits of organic farming

Organic farming considers the intermediate and enduring end product of farming interventions on the agro-ecosystem. Organic farming aims to manufacture food, whereas establishing an ecological equilibrium for prevention of soil fertility and other related problems. This method takes a positive move forward, as opposite to treating the problems when they come into view.

1.4.1. Soil

Soil structure practices such as crop rotations, symbiotic associations, and organic fertilizers are middle to organic practices. These promote soil fauna and flora by improving soil formation and structure. In turn, nutrient and energy cycling is increased and the retentive abilities of the soil for nutrients and water are enhanced, compensating for the non-use of mineral fertilizers. In soil erosion control such management techniques also play an important role. Crop export of nutrients is usually compensated by farm-derived renewable resources, but it is sometimes necessary to supplement organic soils with potassium, phosphate, calcium, magnesium, and trace elements from external sources [6-8].

1.4.2. Air

Organic farming reduces non-renewable energy use by decreasing agrochemical needs. It contributes to mitigating the greenhouse effect and global warming through its ability to appropriate carbon in the soil. Many running practices include recurring yield residues to the soil, use of crop rotations, returning of carbon to the soil for increasing the productivity, and increasing addition of nitrogen-fixing legumes. In many different studies, it was reported that the soils under organic farming have more carbon content as compared to other soils. The more organic carbon is retained in the soil, the more the mitigation potential of agriculture against climate change is higher [9-11].

1.4.3. Water

Pollution of ground water with synthetic fertilizers and pesticides is a major problem in many cultivation areas. Synthetic fertilizers are prohibited in organic farming, they are replaced by compost, animal manure, green manure (organic fertilizers), and through the use of greater biodiversity they contribute to enhance the structure of soil and water infiltration capacity. Risk of ground water pollution may be greatly reduced by properly managed organic systems.

Organic agriculture is greatly expectant as an uplifting measure in those areas where pollution is a genuine dilemma [12].

1.4.4. Genetically modified organisms

The use of these within organic systems is not permitted during any stage of organic food production because their potential impact on health and environment is not entirely understood. Organic farming encourages natural biodiversity. The organic label provides an assurance that these organisms have not been used intentionally in the production and processing of organic products. In conventional farming, increasing the use of genetically modified organism and due to the method of transmission of these organism in the environment (through pollen), organic farming will not be able to ensure that organic products are completely free from genetically modified organism in the future [13-15].

1.4.5. Biological services

The collision of natural farming on usual resources favors connections that are vital for both organic production and nature protection within the agro-ecology. Biological services results include stabilization forming and conditioning of soil, nutrient and waste recycling, predation and habitats. Development of pollution-free agriculture systems depends upon the consumer's purchasing power to buy organic products [6-7].

2. Pollution prevention in organic farming

Getting higher resources of farming and cultivation is why humans live in this world. Farming is an essential resource of continued existence; the lack of these resources leads to famines all over the world. Organic farming was a natural process for the last several years that did not harm the land; many generations of crops have been produced without affecting the fertility of soil. However, modern farming practices have started farming pollution that affects the ecosystem, land, and environment. Farming is a multifaceted activity in which the growth of crops and livestock has to be balanced perfectly [16].

2.1. Causes of farming pollution

2.1.1. Fertilizers

In earlier days, fertilizers have been considered the source of pollution, but in modern days, they treat local pests with new persistent species that have existed for many years and they are loaded with chemicals that are not natural. When pesticides have been sprayed, it mixes with the water and seeps into the ground. Plants absorb the leftover pesticide, and as a result, local streams become contaminated. When these crops are eaten by animals, they are also affected [17].

2.1.2. Livestock

In the past, livestock (cattle, sheep, pigs, chickens) were fed with natural diets, which was supplemented by the waste left over from the crops, and farmers would like to keep them on land. Thus, the animals helped to maintain the farm health as well. But these days, livestock is raised in overcrowded areas, fed with unnatural diets, and sent to slaughterhouses regularly. They cause farming pollution by means of emissions [18].

2.1.3. Weeds and pest

Reducing the natural species and growing unusual crops has become the standard in farming in different areas. The entry of new crops in the local market has resulted in new pest diseases and weeds that the population is not capable of fighting. As a result, local vegetation and wildlife are destroyed permanently. This simply adds to the process of farming pollution [19].

2.1.4. Contaminated water

One source of pollution is the use of contaminated water for irrigation. The water we use comes from ground water reservoirs that are clean and pure water. Other sources are polluted with organic compounds and heavy metals due to the disposal of industrial and agricultural wastes in local bodies of water. As a result, crops are exposed to that water and the process of agricultural pollution becomes harder to fight when such water poisons the livestock and causes crop failure.

2.1.5. Sedimentation

Soil has many layers but only the top layer supports farming. One common reason for the declining soil fertility is inefficient farming practices. Due to these practices, soil left open is eroded by water and wind. This soil is then deposited somewhere and causes sedimentation. This sedimentation causes soil rise in areas such as rivers, streams, ditches, and surrounding fields, and the process of agricultural pollution prevents the natural movement of water, aquatic animals, and nutrients to other fertile areas.

2.2. Effects of farming pollution

2.2.1. Effects on aquatic animals

Organic matter such as ammonia or fertilizers turned into nitrate decreases the level of oxygen in the water and causes the death of many aquatic animals. From animal wastes, bacteria and parasites can get into drinking water, which can cause serious health problems for a variety of aquatic life and animals. It is a hard issue to keep farming pollution in check as it seems. It is difficult to keep track of water levels, soil cleanliness, and industrial pollution. For the last few years, governments have become stricter about enforcing rules. Farmers are becoming aware about the damages and are looking for solutions; most of them are moving toward conventional farming. But for the process of farming pollution to be fully reigned in, there has to be a complete shift in the way cultivation is practiced.

2.2.2. Effects on health

The main source of pollution in water and lakes is farming pollution. Fertilizers and pesticide chemicals are absorbed by ground water and end up in drinking water and cause severe health problems. Oils, degreasing agents, metals, and toxins from farm equipment cause health problems when they get into drinking water.

2.3. Pollution prevention practices

Pollution prevention means reducing the originating of wastes. This will include practices that conserve natural resources by eliminating pollutants through increased efficiency in the use of raw materials, energy, water, and land. Pollution prevention minimizes pollution at the source, so pollution is not created in the first place and never enters into the environment. Environmental prevention has involved controlling and treating the pollution, which in many cases we continue to create. It is helpful in reducing the risks on health and the environment in many ways, such as eliminating the risks associated with the release of pollutants to the environment, avoiding the shift of pollutants from one medium to another medium, and protecting the natural resources for future generations. Pollution prevention can be promoted through several ways such as using voluntary pollution reduction programs, engaging in partnerships, providing technical assistance, funding demonstration projects, and incorporating cost-effective pollution prevention alternatives into regulations. It also involves using systematic management methods such as grass and tree planting technology, improvement of medium and low farmland, and overall use of rural energy resources in order to deal with and improve the ecological environment [20].

3. Management practices in organic farming

In production methods, soil texture plays a bigger role. It influences when a producer can till, the types of tillage methods used, and the frequency of green manure crops. The production methods developed are suited to the climate and soil texture of their farms.

3.1. Healthy soil

In an organic farming system, soil health is the key to success. Soil health can be assessed qualitatively. Many producers look for a dark, rich-colored soil with earthy smell and good organic matter. Earthy smell indicates that the soil is rich of microorganisms, which are vital to soil health. Some take a note of wildlife attraction to the field; birds can be a good indication of earthworms and other organisms. Some producers note the color of leaves and the development of root systems as crops grow; yellow leaves indicate low nitrogen levels, red color and dead spots indicate a plant is under stress, and dark green color with slow growth indicates low nutrient levels. Weeds growing in the field indicate which nutrients are available in the soil; they require the same nutrients but in different amounts. Fertile soil is called healthy soil; it contains sufficient chemical nutrients (macronutrients and micronutrients) for plant growth.

Those needed in larger amounts are called macronutrients such as nitrogen, phosphorus, calcium, sulfur, and potassium. Among them, nitrogen is commonly limited to plants and it is abundant in air; few free-living microbes and rhizobium associated with legumes can fix the nitrogen from air. While other minerals can move into the soil from the underlying rocks. When products are removed from the farm ecosystem, nutrients are removed from the soil. Among them, nitrogen is removed in the largest quantities, but fortunately it can be replaced from the air. Fertile soils can be easily tilled and have good structure, it allows good penetration and absorption of nutrients. Biological fertility such as microbes cycle chemical nutrients available via the breakdown of plant residue and animal wastes. They form a symbiotic relationship with the plants that increase the amount of soil that plants are able to search the nutrients.

3.1.1. Soil test

To check the level of soil fertility and nutrients, soil test may be needed. Soil test provide information about soil nutrients, pH, and organic matter. Some soil test results include macronutrients. These soil tests typically provide recommendations about fertilizers in farming. Soil testing can be beneficial for organic producers. Long-term changes in soil fertility help the producers to adjust soil management strategies such as crop selection, rotation, and green manure. Experienced producers do not feel the need to test the soil; they evaluate the health of the soil using production yield. For the soil test, it is very important that soil samples be collected and stored properly according to the instructions of the laboratory, especially in the case of soil biology, as soil organisms can die or multiply rapidly and this may invalidate the results. A few soil tests that are used by organic producers are as follows:

1. Soil food web Canada, Inc., measures the biodiversity (quantity of bacteria, fungi, and nematodes) in the soil, suggests optimal levels for different crops, and provides suggestions to increase the activity of soil.

2. Western Ag Innovations Inc. evaluates soil fertility by using a Plant Root Simulator probe. For this purpose, probes are placed in the soil for different time periods and measure the level of nutrient across the membrane. It will give a good estimate of nutrients available to the plants.

3. Kinsey's Agricultural Services analyze the soil sample by using the Albrecht system. Their recommendations are based on fertilizer preference, crop history, and type of operations.

4. ALS Laboratory group assesses the level of macronutrients and micronutrients in the soil. This test measures the level of nutrients that can be extracted, including organic matter, pH, and cation exchange capacity.

3.1.2. Soil biology

Soil biology can be encouraged by several methods. Many experienced producers suggest that green manure is one of the best methods to maintain the life of soil; other methods are animal manure and straw residue, selecting good rotation, and reducing tillage. Many farmers recommended that all straw be worked back into the soil to return the nutrients. They provide microorganism to increase the organic matter of the soil. Legume incorporation causes a change in microbial population toward greater metabolic activity and increases organic matter. Soil microorganisms are also affected by tillage; mostly producers try to keep less tillage operation and maintain some cover on all fields throughout the growing season. For this purpose, green manure is the best strategy as it covers the land and protects it from drying out. Organic producers must care and try to avoid methods that increase the soil erosion and kill soil microbes [21].

3.1.3. Soil organic matter

Organic matter is the key for maintaining water holding capability and soil health. Animal and plant residue, along with the soil organisms such as bacteria, fungi and nematodes, are the component of organic matter remains in the soil worked from year to year. As a result of climate and vegetation that existed before the land was broken, organic matter is formed. The four different divisions of soil organic matter are fresh organic matter, decomposing organic matter, stable organic matter, and living organism. When fresh organic plant material is added to the soil, microbes break it down and this moderately decomposed organic matter holds nutrients for growing plants. In the decomposition process, stabilized organic matter is the final product; it provides structure to the soil resulting in good aeration and water holding ability [22].

3.1.4. Soil applied

Some experienced producers use calcium, sulfur, gypsum, and rock phosphate after soil tests indicate low levels of nutrients. To improve the soil biology, microbial organisms are also used.

3.1.5. Foliar applied inputs

Some producers use foliar sprays as inputs on the plant when it is growing. These can be used to control the disease or to reduce the risk of disease. Most often, the intention is to feed the helpful organisms that reduce the risk of pathogens.

3.1.6. Manure and compost

Manure is an excellent organic fertilizer; its use is highly regulated by organic standards. To build the soil fertility many livestock producers use it. It can be used in different forms such as organic composted manure, deposition on crop land, and application of manure without being composted. For proper decomposition, it should be applied at a suitable time of the year and at a proper peak in the rotation. For more effectiveness, fresh manure should be incorporated soon to decrease the nitrogen loss and it should be applied in cool conditions; however, many producers will age manure for several years before putting it in the field, which is not so good [23].

Composting is one step forward to manure; it is a process that can be described as the aerobic decomposition of organic matter to produce a humus-like product called compost. In this process, microorganisms (fungi) are involved that convert the manure to humus, which is darker in color and has an earthy smell. Composting requires some machinery and effort to maximize the humus-producing potential of manure. To meet compost standards, producers must mange proper air, moisture, and temperature in the mass. Proper composting balance between carbon and nitrogen proportion is necessary. Careful planning is required when making an allowance for compositing animal wastes on farms. The location of the compost site matters a lot to avoid risks to ground water and nearby water sources. Enclosing livestock and collecting, transporting, and spreading compost and manure are costly and inefficient. The simple method adopted by some producers is that they allow livestock to graze crop land and put the fertilizers straight onto the field [24].

3.1.7. Nutrient amendments

A few producers use amendments such as seed inoculants and foliar spray on the green parts or soil. Organic amendments have very little reliable use. These products should be used carefully. Before using any amendment, it must be ensured that it is approved for organic production.

3.1.8. Seed inputs

Nitrogen fixation is very important for plant growth. For nitrogen fixation, rhizobial inoculants with legumes are used as they create an environment that favors the bacteria responsible for nitrogen fixation. They do not need seed inoculants, but some experienced producers suggest that if the inoculant is applied on or below the seed they give better results. Some additional products such as humates, mycorrrhizal fungi, and other microbes respond to crops differently, their response depending upon crop, crop cultivar, and management history.

3.1.9. Green manures

A green manure is a crop worked into the soil to provide nutrients to the organisms and ultimately to the crops. To sustain a healthy soil, the use of green manure in crop rotation plays an important role. Green manure is a legume that fixes nitrogen into the soil; availability of nitrogen depends on the growing condition, moisture, and inoculation. Producers recommended sweet clover, alfalfa, red clover, field pea, and faba bean for nitrogen fixation and oilseed and buckwheat to improve phosphorus availability.

3.1.10. Rotation of crops

Rotation is a planned sequence of crops, and organic producers consider it as the most important key in organic farming. A lot of scientific literature suggests that crop rotation is more beneficial than monocultures. The more variable the rotation, the more stable the yield. Resources can be used more effectively by rotating the crops with different characteristics. As we know, crops differ in their requirements of water, nutrients, and susceptibility to pests and diseases. The sequence of crops must be cautiously selected, which is well adapted to the fertility level, to avoid the disease potential that builds in crops. Rotation is planned according to the health of the soil such that crops that require tillage should be balanced with crops that build organic matter, and crops that utilize more nitrogen should be balanced with crops that supply nitrogen. Crop rotation is also very important in weed management. For different crops, different weed management practices are used. Each type of management practice is a disturbance that favors one weed species over the other. If annual crops are rotated with winter crops, then the disturbance pattern is varied and different species are disadvantaged at different times. This results in a more diverse weed community. This diversity can be beneficial as it increases the variety of food and shelter available to the beneficial organisms. Rotations are also crucial to insects and disease management. Insects and diseases are specific to a single crop; if they remain away for a long time from that crop, they are not able to increase to a dramatic level. Most producers consider rotation to be a work in progress that will change as the soil changes. A flexible rotation is recommended by most experienced producers to respond to changes in disease pressure, market, and contaminations by microbes. Organic producers take soil samples every couple of years and spend time in learning how they can improve the farming techniques.

3.2. Seeding

Seeding is the time when planning and reality come together. Most often, the weather determines when to seed, what to seed, and which equipment to use. The ideal time for seeding is when it grows in a weed-free environment. Weeds are much more competitive when the crop emerges, as compared to the established crop. Mostly producers are not able to buy new equipment for seeding; it is time to consider what can be done to give the best advantage to the crop. The time of seeding is very important; in wet years you can seed anytime, but in dry years, you must seed as early as possible. Try to avoid seeding in very hot temperatures; if you want to seed early, then notice the condition of soil; if you seed late, then control the weeds.

A number of factors are considered by producers for crop selection such as soil fertility, weed control, crop type, and previous crop in rotation, but experienced producers follow some criteria and then choose the variety of crop to plant. In this criteria, the varieties to select from are based on which ones grow well and have disease resistance, heritage varieties, high-quality crops such as wheat with high protein, varieties that are in demand in organic markets, and varieties that can give viable seeds for the next year. Producers identify the characteristics that are best suited to the organic production and then they seed. Organic producers think that heritage varieties are best because they are developed without chemical and fertilizer inputs. Under organic management, producers can perform and yield well.

Seed quality and seeding rate are very important in organic farming. Experienced producers do not consider it necessary to use certified seeds; some suggest it is important only when it was time to renew the seed. Some scientific studies confirm the advantage of high seeding rate. Higher seeding rate can increase the crops' ability to cover land. Increasing seeding rate may be more important under conditions of higher fertility, when weeds may be more competitive. Crop emergence can be affected by seeding equipment. Organic farmers favor different types of seeding equipment. The most preferred seeding equipment that are used by organic producers are air seeder, disk seeder, double disk press drill, and valmar spreader. These are used for different seeds according to the climatic conditions.

3.3. Weeds

For new organic farmers, weed management is very threatening. Organic fields share the same weeds as other farms. For determining the weed community, some factors are important such as soil texture, environmental variables, and crop rotation. The most common weeds, such as wild oats, bluebur, stinkweed and wild buckwheat, are found on organic farms. Many producers suggest that tillage can be a powerful weed management tool especially before and during seeding. Weeds that emerge before the crop gain more of the resources and thus have much more effect on the crop than weeds that emerge later. A second option for weed control exists after seeding but before the crop emerges. Some successful weed control practices used by organic producers are the use of solid crop rotation, delayed seeding, seeding with high rate, spiking in the fall to control quack grass, and growing alfalfa and sweet clover to suppress weeds. It is also very important in weed management practices to know the ecology of weed management.

3.3.1. Ecological weed management

For weed management, most of the organic farmers rely on multiple plans. Ecological weed management promotes weed suppression, instead of weed elimination, by increasing crop competition and phytotoxic effects on weeds. A specific method such as crop rotation is one of the best methods used by organic farmers to control weed management. Organic producers suggest that small grains or legumes must be planted for at least one year out of every five years to maintain soil health. If the legume is plowed under as a cover crop in the fifth year, four years of row crops may be grown prior to the green manure crop year. The same crop cannot be grown in sequential years; due to this, soybean cannot be grown in the same field year after year. The ideal crop preceding soybeans is winter rye. Soybean fields are rotated to a small grain (oats, barley, wheat, or rye) or corn.

3.3.2. Production practices

Organic farmers suggest some production practices for weed management such as variety selection where farmers select crop varieties (e.g., quick canopy-forming) that compete well with weeds within and between rows. As regards crop density, planting at the utmost modified population will provide the crop an enhanced competitive border over weeds. Closer row spacing generally has greater crop competition with weeds in row middles. For the rapid canopy, high germination rate seeds are more preferable. Date of sowing matters a lot; warm season crops are planted when the soil is warmed properly to facilitate the germination.

3.3.3. Physical tactics for weed management

These are the key factors to control weed management on all organic farms; it includes mulching, cultivation, and propane flame burning. Mulching is used in combination with manual labor in many horticulture operations for proper weed control. It is of two types, natural and synthetic mulches. These are used in organic operations along with polyethylene film and polypropylene landscape fabric. Mulch can be made from small grain and soybean straw. During decomposition, organic mulches add organic matter to increase soil porosity, water holding capacity, microbial populations, and cation exchange capacity. Straw mulch is used in organic horticultural operations, for example garlic, strawberry, and herb farms, to control weeds and protection from harsh environments.

Timely cultivation is critical in organic weed management. Depending on the crop, cultivation offers the least labor-intensive weed control method. Midwestern organic farmers used two to three row cultivations. First cultivation occurs at a slow speed, second cultivation usually is completed at mid-season at a faster speed, while third cultivation is again performed at a slow speed. Propane flame-burners have been added as an additional tool in their weed management toolbox by many organic farmers. When tillage with large machinery is not feasible, flaming is used during high field moisture, while in drier weather it is used in conjugation with cultivation.

3.4. Insects

The most experienced organic producers are serious about insect problems. During dry season the most common and problematic insect is the grasshopper. If the crop and soil were healthy, then there would be less insect problems. There are some specific recommendations for insects: for grasshoppers, use tillage to avoid egg laying, use foliar sprays, seed early and use alfalfa border; for lygus bugs, delay seeding; for wheat midge, select resistance varieties and delay seeding; and for aphids, keep an environment where predators flourish. Most producers do the best they can to control insects.

3.5. Tillage

In organic and conventional farming, soil erosion resulting from tillage is a major concern. Organic producers use more tillage; they use it for seed bed preparation, weed suppression, and for the incorporation of green manure. To prepare seed bed and to control weeds tillage, a harrow and cultivator is used; those who used disc seeders reported less cultivation because this method killed weeds. Some farmers used light tillage with harrow to control weeds before and after crop emergence. In a survey, organic producers were asked about the increase or decrease in their tillage operations, they replied that type of tillage had changed. At the beginning, organic producers used more tillage operations to control the weeds, but after that producers moved toward less tillage.

Tillage can be reduced, although it is the only method of terminating the green manure. Recently, producers have challenged the belief that tillage is needed for green manure termination through the method of rolling, mowing, and blading. One producer indicated that the wide-blade cultivator causes minimal disturbance to the soil and leaves much residue, so this was an effective way to terminate green manures while reducing the risk of soil erosion. Generally, tillage operations are used more in black soil where there is high weed pressure. Producers who used more tillage operations try to minimize erosion potential by understanding the condition of the soil.

3.6. Transition

A transition from conventional to organic farming is not an easy step; it takes time and requires a change in mind set. Some producers suggest that transition in the mind takes longer compared to the transition on the land. Producers learn more because new methods have come into practice such as green manure, rotation of crops, mechanical weed control, organic fertility management, and erosion reduction. Transition time is very important, because it provides time for the soil to become free from chemicals that remain in the soil due to conventional farming. Weed control and soil fertility is the top priority. It is an economically vulnerable time. Although the transition is difficult, organic farming made them feel empowered. New organic farmers recommend the following about tillage during the transition years: understand the soil; till in different directions in different years; for weed control, keep tillage to the minimum need; replace black fallow with weed fallow; try to avoid tilling light soils in dry years; and harrow the cereals when they are about four inches in height.

4. Conservation in organic farming

The most important key for sustainable cultivation is the conservation of natural resources, especially considering the decreasing conditional subsidies of the Common Agricultural Policy of the European Union for the coming years. If lower economic support compels farms to increase efficiency to reduce production costs, at the same time providings an interaction of agricultural activities with environment quality, suitable natural resources management will be a vital feature for farms.

4.1. Soil conservation

Soil is the production base of all agricultural systems and its conservation is the pillar of sustainability. Soil quality is affected by wind and water erosion and farming practices. Soil erosion is one of the factors of organic farming, so it is necessary to develop soil conservation practices. Conservation practices are usually those that decrease wind speed, reduce rate and amount of water movement, and raise soil organic matter levels. All these conservation managements are not employed to all situations; the management will depend upon the soil type, climate, topography, and type of farming in that area. Producers can use a number of conservation practices that are best for their farms. However, organic crop producers have to face great challenges because conservation practices that use herbicides are not an option. Some common organic crop production practices, such as post-emergent harrowing for weed control, are destructive to the soil. So producers may need to employ some additional conservation measures if practices such as post-emergent harrowing are used. To conserve the soil, some strategies are presented as follows.

Crop residues (roots, chaff, stems, and leaves) are the key source of organic matter replacement. These residues also contain nutrients such as phosphorus, sulfur, potassium, nitrogen, and micronutrients. They improve soil properties such as water infiltration, water storage, and particle aggregation. Among crops, the amount of residue produced and the rate of decay are

Soil degradation processes		Soil conservation practices
Soil erosion Crusting Compaction Waterlogging Acidification Desrtification Nutrient runoff	**Soil** — **Productivity** +	Crop rotation Contour farming Chemical fertilizers Tillage conservation Water conservation Residue management

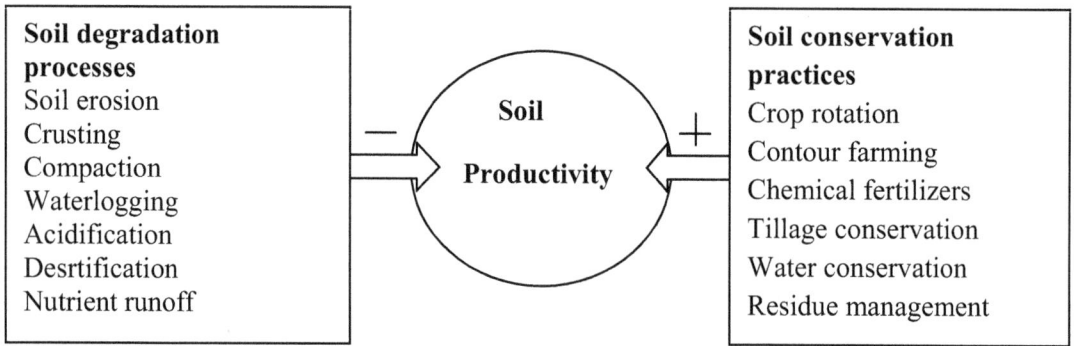

different. The combination of these two factors determines the quality of residue in relation to its value for soil conservation [25].

4.1.1. Forage crops

Forage crops contribute significant amounts of organic material to the soil and offer an alternative product in the form of hay, silage, or seed. Forage production for two to four years should also be considered as part of a normal crop rotation. Selection of forage species and management practices can be customized to specific problems such as drought, salinity, poor soil structure, low pH, and excessive soil moisture.

4.1.2. Stubble cutting

Moisture conservation is also important because the additional moisture will improve crop growth. It may also allow extending the rotation, which is another conservation practice. It can be enhanced by trapping more overwinter snow with "tall" or "sculptured" stubble. Tall stubble refers to stubble that is cut 12 inches high, while sculptured stubble refers to alternate swaths that are cut at normal height and taller.

4.1.3. Direct seeding

In organic farming, herbicides cannot be used. Organic crop production is not usually associated with direct seeding but some producers do put into practice direct seeding. However, organic producers possibly will think about this protection practice when low weed pressure and previous crop straw and chaff have sufficiently spread [26].

4.1.4. Balancing of rotational crops

An ideal rotation should be as diverse as possible; a diverse crop rotation can help soil nutrient availability because different crops remove different nutrients. Most commonly, sixteen essential nutrients are present in soil. In the rotation, growing legumes provides both nitrogen and non-nitrogen benefits to following crops. If legumes are inoculated properly, they fix 90% of their nitrogen necessity from the air and rest is obtained from the soil. However, during the growing season, nitrogen is exuded from legume roots and the legume residue decomposes and recycles the nutrients quicker than non-legume residues, thus more nitrogen is regularly accessible to the following crop than if a non-legume had been grown. When planning complementary rotational cropping, growth patterns of a variety of crops should also be taken into account. Crops with broad leaves such as polish canola, lentil, flax, and pea take out nutrients and moisture from more shallow rock bottom than cereals that belong to spring-seeded. Thus, winter wheat rooted deep uses moisture in the early growing season while the recurrent forages use nutrients and moisture from subsoil because they are deep rooted. Shallow-rooted crops are best adapted as compared to deep roots because they will not expand energy in search of moisture as compared to other crops. Medium root crops come into view as enhanced and modified to pursue shallow-rooted crops as they benefit from any moisture left at the depth, which is not used by the preceding shallow-rooted crop [27].

4.1.5. Total crop rotations

Summer fallowing is destructive to the soil because no new organic matter is returned to the soil during this year. Breakdown of soil organic matter increases due to tillage. Extending crop rotations is a conservation practice because it reduces the incidence of summer fallow. This practice can improve fertility, collective constancy, tilth, damp storage space, and conflict to soil erosion and deprivation, in addition to decreasing insects and disease problems. All these reported factors enhance yield productivity and have positive effects on soil sustainability. Decisions for cropping strategies would not be for a short duration but the long-term effects on the soil and environment should also be considered. A varied crop rotation should comprise pulses, seed oil, fall-seeded crops, and forages. Crop diversity level determines the implication of the rotational payback. During rotation, some selection and management of legume species is a vital aspect of achieving diversity and supplying nitrogen through symbiotic nitrogen fixation.

4.1.6. Tillage

During tillage crop residue, conservation is affected by the equipment type, speed, depth, and frequency of tillage, as well as soil and climatic factors. Limiting all these factors conserves crop residue and soil moisture. It has been difficult to convince researchers and extension services that rigorous tillage does not allow for soil and water conservation and decreases soil natural content. Tillage may be defined according to conservation farming as the integration of agronomic practices with the aim of conserving, improving, and efficiently using natural possessions [28]. On yield consistency, the farmers' point is correct but the reason of low yield in conservation tillage systems is only associated with the first few years of the changeover period between conservation practices and intensive tillage. Energy can be saved by adopting the method of reduced tillage and greater savings can be achieved by no-tillage [29]. Greater benefits can also be noticed in relation to environmental aspects; large amounts of crop residues on the soil surface reduce water runoff and nutrients loss [30].

In tillage operations during shallow tillage, crop residue accumulates near the soil surface and it will be most effective in reducing wind and water erosion by improving infiltration and reducing evaporation. Reducing tillage speed generally reduces crop residue burial. Residue conservation is significantly influenced by tillage equipment type; for instance, a wide-blade cultivator preserves considerably more remainder than a cultivator that is considered better than a discer. The addition of harrows to a field increases the amount of remnants buried, while adding a rod weeder to a cultivator does not considerably affect deposit lessening. The need of each tillage operation should be carefully considered according to the type of soil, but tillage should be avoid under wet soil conditions as this can degrade soil structure and significantly decrease surface residue levels [31].

4.1.7. Wind barriers

4.1.7.1. Annual crop barriers in crops

Taller annual crops have been used as barriers to a restricted degree in low residue-producing crops. A divider is placed in the seedbox so that two rows of wheat are seeded every seeder width. At harvest, the lentil is combined and the barrier strip left standing to trap snow and prevent wind erosion during the upcoming winter.

4.1.7.2. Strip cropping

In strip cropping, alternating strips of crop and summerfallow consists at an angle perpendicular to the prevailing winds. According to the soil, texture strip width varies. Wind erosion is more common in sandy soils as compared with clay and loam soils. Strip cropping works well for loam and clay soils where increase stripping will considerably decrease the potential for wind erosion. In sandy soil types, too many strips are essential to be convenient. When establishing strip widths, the size of field equipment should be kept in mind. This practice is more common in drier areas; however, it can be used in wetter areas where the pattern of strip formation is changed to avoid water erosion.

4.1.7.3. Cover crops

Rotations should also comprise the use of cover crops to protect the soil from water and wind erosion throughout susceptible periods, for instance, summer fallow when normally position stubble does not exist. Cereal yield should be seeded between August and September; fall frosts will be able to kill plant material and remain on the soil surface until spring planting, providing valuable soil protection. Winter wheat, fall seeded cereal may be used in a parallel fashion. In the following spring, these crops may be removed by tillage or used for short-term livestock grazing, or grown to maturity in the case of winter wheat or fall rye [32].

4.1.7.4. Shelterbelts

It can effectively decrease wind velocity for a distance of 20 or more times than their height. They effectively control wind erosion when planted at the right angles to current winds. The effectiveness of shelterbelts depends upon maintenance, in addition to height. They may also be helpful for increasing crop yields.

4.1.7.5. Perennial grass barriers

These are two rows of grass planted at right angles to current winds to decrease wind erosion, entrap snow, and reduce evaporative losses. Placement of barriers depends on soil type; these are closest on sands, moderately spaced on clays, and utmost apart on loams. Barriers may be placed further apart if other soil conservation practices are also being used. Tall wheatgrass is a weak participant with most field crops and will not spread beyond the seeded rows. It also grows high enough without accommodation to trap snow, helping in soil moisture renewal.

4.1.8. Green manure

The assimilation of any green vegetative material into the soil is called green manure. In crops, it adds organic matter to the soil and improves soil health. The extent of soil improvement depends on the type and quantity of plant material returned to the soil. Biennial or perennial legumes as green manure give great benefits to soil with poor level of organic matter but the time of implanting these legumes matters a lot. Grain legumes, such as pulses, can be used as green manure effectively because their annual growth habit will not contribute in nitrogen fixation as biennial or perennial legumes. However, they are more flexible to an accessible crop rotation. Non-legume crops can also be used as a green manure crop [33].

4.1.9. Animal manure

Animal manure, such as livestock and poultry, provides not only nutrients to plants but also affects soil tilth and particle aggregation. Organic matter contained in manure act as binding agents in stabilizing soil structure. The addition of manure changes the soil structure and this surely affects water infiltration, water holding capacity, and aeration, as well as resistance to wind and water erosion. Manure nutrient value depends upon some factors such as animal type and age, type of feed, amount of straw, and method and time of storage. In the manure, some micronutrients are helpful to prevent the plant deficiency symptoms from happening. The rate of manure application recommended by different soil testing laboratories that test the animal manure for nutrient content depends upon the availability of soil type, slope, location, and different construction practices. For the prevention of environmental contamination, rates of manure application should not exceed what a crop can use in one growing season. Following manure application to prevent nitrogen loss, it should be incorporated as quickly as possible into the soil for proper plant growth [34].

Acknowledgements

This research is supported by UMRG (RG257-13AFR) IPPP (PG038-2013B) and FRGS (FP038-2013B).

Author details

Maliha Sarfraz[1], Mushtaq Ahmad[2], Wan Syaidatul Aqma Wan Mohd Noor[3] and Muhammad Aqeel Ashraf[4,5*]

*Address all correspondence to: chemaqeel@gmail.com

1 Institute of Pharmacy, Physiology & Pharmacology, University of Agriculture Faisalabad, Pakistan

2 Department of Plant Sciences, Quaid-i-Azam University Islamabad, Pakistan

3 School of Biosciences and Biotechnology, Faculty of Science and Technology, Universiti Kebangsaan Malaysia (UKM) Bangi, Selangor, Malaysia

4 Department of Environmental Science and Engineering, School of Environmental Studies, China University of Geosciences, Wuhan, P. R. China

5 Water Research Unit, Faculty of Science and Natural Resources, University Malaysia Sabah, Kota Kinabalu, Sabah, Malaysia

The authors certify that there is no conflict of interest with any financial organization regarding the material discussed in the paper.

References

[1] Kiyani, S., Ahmad, M., Zafar, M. A., Sultana, S., Khan, M. P. Z., Ashraf, M. A., Hussain, J. & Yaseen, G. 2014. Ethnobotanical uses Of medicinal plants for respiratory disorders among the inhabitants Of Gallies-Abbottabad, Northern Pakistan. Journal of Ethnopharmacology, 156, 47-60. DOI: 10.1016/j.jep.2014.08.005.

[2] Naureen, R., Tariq, M., Yusoff, I., Choudhury, A. J. K. & Ashraf, M. A. 2014. Synthesis, spectroscopic and chromatographic studies of sunflower oil biodiesel using optimized base catalyzed methanolysis. Saudi Journal of Biological Sciences, 22, 322-339. DOI:10.1016/j.sjbs.2014.11.017.

[3] National Organic Standards Board (NOSB) 1995. Definition of organic. Drafted and passed at April 1995 meeting in Orlando, FL. 1995.

[4] Qureshi, T., Memon, N., Memon, S. Q. & Ashraf, M. A. 2015. Decontamination of ofloxacin: Optimization of removal process onto sawdust using response surface methodology. Desalination and Water Treatment, 1-9. DOI: 10.1080/19443994.2015.1006825.

[5] Reaganold, J., Elliot, P. & Unger, Y.L. 1987. Long-term effects of organic and conventional farming on soil erosion. Nature, 330, 2-10. DOI: 10.1038/330370a0.

[6] Reicosky, D. C. 2001. Conservation Agriculture: Global Environmental benefits of soil carbon management. In: Conservation Agriculture, a Worldwide Challenge (Garcia Torres L; Benites J; Martinez A, eds), pp 1–12. Kluwer Academic Publisher, Dordrecht, The Netherlands.

[7] Ahmed, Q., Yousuf, F., Sarfraz, M., Mohammad Q.A., Balkhour, M., Safi, S.Z. & Ashraf, M.A. (2015). Euthynnus affinis (little tuna): Fishery, bionomics, seasonal elemental variations, health risk assessment and conservational management. Frontiers in Life Science, 8 (1), 71-96.

[8] Borin, M., Menini, C. & Sartori, L. 1997. Effects of tillage systems on energy and carbon balance in North-Eastern Italy. Soil and Tillage Research, 40, 209-226

[9] Butt, M. A., Ahmad, M., Fatima, A., Sultana, S., Zafar, M., Yaseen, G., Ashraf, M.A., Shinwari, Z. K. & Kayani, S. 2015. Ethnomedicinal uses of plants for the treatment of snake and scorpion bite in Northern Pakistan. Journal of Ethnopharmacology, 1, 1-14. DOI:10.1016/j.jep.2015.03.045.

[10] Drinkwater, L.E., Wagoner, P. & Sarrantonio, M. 1998. Legume-based cropping systems have reduced carbon and nitrogen losses. Nature, 396, 4-16. DOI: 10.1038/24376.

[11] Gharibreza, M., Raj, J. K., Yusoff, I., Othman, Z., Zakaria, W., Tahir, W. M. & Ashraf, M. A. 2013. Historical variations of Bera Lake (Malaysia) sediments geochemistry using radioisotopes and sediment quality indices, Journal of Radioanalytical and Nuclear Chemistry, 295(3), 1715-1730.

[12] Ahmed, Q., Yousuf, F., Sarfraz, M., Bakar, N. K. A., Balkhour, M. A. & Ashraf, M. A., (2014). Seasonal elemental variations of Fe, Mn, Cu and Zn and conservational management of Rastrelliger kanagurta fish from Karachi fish harbour, Pakistan. Journal of Food, Agriculture and Environment, 12 (3&4), 405-414.

[13] Ashraf, M. A., Ahmad, M., Aqib, S., Balkhair, K. S. & Bakar, N. K. A. 2014. Chemical species of metallic elements in the aquatic environment of an ex-mining catchment, Water Environment Research, 86 (8), 77-728.

[14] Ashraf, M. A., Yusoff, I., Yusof, I. & Alias, Y. 2013a. Study of contaminant transport at an open tipping waste disposal site. Environmental Science and Pollution Research, 20 (7), 4689-4710.

[15] Ashraf, M. A., Ullah, S., Ahmad, I., Qureshi, A. K., Balkhair, K. S. & Rehman, M. A. 2013b. Green Biocides, A Promising Technology: Current and Future Applications, Journal of the Science of Food and Agriculture, 94 (3): 388-403. DOI: 10.1002/jsfa.6371.

[16] SSSA 1997. Glossary of Soil Science Terms. Soil Science Society of America, Madison, WI p 134.

[17] Ashraf, M. A., Maah, M. J. & Yusoff, I. 2012d. Chemical speciation and potential mobility of heavy metals in soil of former tin mining catchment, The Scientific World Journal, 125608, 1-11

[18] Ashraf, M. A., Maah, M. J. & Yusoff, I. 2012e. Chemical speciation of heavy metals in surface waters of former tin mining catchment, Chemical Speciation and Bioavailability, 24 (1), 1-12.

[19] Ashraf, M. A., Maah, M. J. & Yusoff, I. 2012f. Bioaccumulation of heavy metals in fish species collected from former tin mining catchment. International Journal of environmental Research, 6 (1), 209-218

[20] Ashraf, M. A., Maah, M. J. & Yusoff, I. 2011a. Analysis of physio-chemical parameters and distribution of heavy metals in soil and water of ex-mining area of Bestari Jaya, Peninsular Malaysia. Asian Journal of Chemistry, 23 (8), 3493-3499.

[21] Ashraf, M. A., Maah, M. J. & Yusoff, I. 2011b. Assessment of heavy metals in the fish samples of mined out ponds Bestari Jaya, Peninsular Malaysia. Proceedings of the Indian National Science Academy, 77(1), 57-67.

[22] Ashraf, M. A., Maah, M. J. & Yusoff, I. 2010. Water quality characterization of Varsity lake, University of Malaya, Kuala Lumpur Malaysia. E Journal of Chemistry, 7 (S1), S245-S254.

[23] Bakar, A. F. A., Yusoff, I., Fatt, N. T., Othman, F. & Ashraf, M. A. 2013. Arsenic, zinc and aluminum removal from gold mine wastewater effluents and accumulation by submerged aquatic plants (Cabomba piauhyensis, Egeria densa, and Hydrilla verticillata). BioMed Research International, 890803, 1-7.

[24] Zulkifley, M. T. M., Fatt, N. T., Raj, J. K., Hashim, R. & Ashraf, M. A. 2014b. The effects of lateral variation in vegetation and basin dome shape on a tropical lowland stabilization in the Kota Samarahan-Asajaya area, West Sarawak, Malaysia. Acta Geologica Sinica, 88 (3), 894-914.

[25] Ashraf, M. A., Maah, M. J. & Yusoff, I. 2012a. Study of chemical forms of heavy metals collected from the sediments of tin mining catchment. Chemical Speciation and Bioavailability, 24(3), 183-196.

[26] Ashraf, M. A., Maah, M. J. & Yusoff, I. 2012b. Chemical speciation of heavy metals in sediments of former tin mining catchment, Iranian Journal of Science and Technology, 36 (A2), 163-180.

[27] Ashraf, M.A., Maah, M.J. & Yusoff, I. 2012c. Morphology, geology and water quality assessment of former tin mining catchment. The Scientific World Journal, 369206, 1-15.

[28] Batool, S., Khalid, A., Chowdury, A. J. K., Sarfraz, M., Balkhair, K. S. & Ashraf, M. A., 2015. Impacts of azo dye on ammonium oxidation process and ammonia oxidizing soil bacteria. RSC Advances, 5: 34812-34820. DOI: 10.1039/C5RA03768A.

[29] Khaskheli, A. A., Talpur, F. N., Ashraf, M. A., Cebeci, A., Jawaid, S. & Afridi, H. I. 2015. Monitoring the Rhizopus oryzae lipase catalyzed hydrolysis of castor oil by ATR-FTIR spectroscopy. Journal of Molecular Catalysis B: Enzymatic, 113, 56-61. DOI:10.1016/j.molcatb.2015.01.002.

[30] Bakar, A. F. A., Yusoff, I., Fatt, N. T. & Ashraf, M. A. 2014. Cumulative impacts of dissolved ionic metals on the chemical characteristics of river water affected by alkaline mine drainage from the Kuala Lipis gold mine, Pahang, Malaysia. Chemistry and Ecology, 13 (1), 22-33.

[31] Siegrist, S., Schuab, D. & Pfiffner, L. 1998. Does organic agriculture reduce soil erodibility? The results of a long-term field study on loess in Switzerland. Environment and Resources Economics, 1998. 69:64. DOI: 10.1016/S0167-8809(98)00113-3.

[32] Surhio, M. A., Talpur, F. N., Nizamani, S. M., Amin, F., Bong, C. W., Lee, C. W., Ashraf, M. A. & Shahid, M. R. 2014. Complete degradation of dimethyl phthalate by biochemical cooperation of the Bacillus thuringiensis strain isolated from cotton field soil. RSC Advances, 4, 55960-55966.

[33] Yusoff, I., Alias, Y., Yusof, M. & Ashraf, M. A. 2013. Assessment of pollutants migration at Ampar Tenang landfill site, Selangor, Malaysia. ScienceAsia, 39, 392–409.

[34] Zulkifley, M. T. M., Ng, N. T., Abdullah, W. H., Raj, J. K., Ghani, A. A., Shuib, M. K. & Ashraf, M. A. 2014a. Geochemical characteristics of a tropical lowland peat dome in the Kota Samarahan-Asajaya area, West Sarawak, Malaysia, Environmental Earth Sciences, 73 (4), 1443-1458. DOI 10.1007/s12665-014-3504-2.

6

Preliminary Results Regarding the Use of Interspecific Hybridization of Sunflower with *Helianthus argophyllus* for Obtaining New Hybrids with Drought Tolerance, Adapted to Organic Farming

Florentina Sauca and Catalin Lazar

Additional information is available at the end of the chapter

Abstract

Taking into account the climatic changes expected in the future, significant shrinking of the current favourable ecological zones for sunflower is anticipated, and the transition period to that situation may be very short. The classical breeding process has a relatively long duration (7-9 years), so breeders are interested in taking advantage of some biotechnological methods (*embryo rescue*) for obtaining new sunflower lines with increasing tolerance to a certain stress factor.

Improving drought tolerance of sunflower cultivars is a priority for a breeding program of the National Agricultural Research and Development Fundulea (NARDI-Fundulea) because it provides stable productions under a changing climate condition already seen in the past twenty years.

In the period between 2008 and 2014 at NARDI-Fundulea, a research project was started to obtain new genotypes of sunflower with improved resistance to drought and heat through interspecific hybridization between *H. annuus* and *H. argophyllus* and that are suitable for application in organic culture. This research project received funding from the World Bank through a MAKIS project.

Keywords: Embryo rescue, interspecific hybridization, *H. argophyllus*, *H. annuus*, NARDI Fundulea, organic farming, drought

1. Introduction

In Romania, Vrânceanu (2000) [1] was able to obtain interspecific progenies *(H. annuus x H. argophyllus)* with drought resistance.

Interspecific hybridization is an additional technique to create new sources of genetic variability for the improvement of sunflower (Christov, 2013) [2]. With all the difficulties that may arise due to differences in the number of chromosomes (2x, 4x, 6x) and crossing incompatibility, interspecific hybridization is considered as an accessible way to incorporate wild germplasm into cultivated sunflower, especially to increase the resistance to abiotic stress factors (Iouraş and Voinescu, 1984) [3].

At the beginning of the project, 27 *H. annuus* parental lines were crossed with *H. argophyllus*, and two generations of interspecific hybrids/year were obtained in the greenhouse and house vegetation of NARDI-Fundulea in the first 2 years after the start of the project.

From each line hybrid obtained in 2008-2009 (Saucă et al., 2010) [4], six plants were selected, and their seeds underwent parallel backcross, self-pollination, and selection procedure.

As a result of this process, seven lines with significantly improved resistance to drought and heat (tested in field and laboratory) and that are suitable for organic farming system were selected in backcross 7. In 2015, these seven uniform lines with high production potential, oil content of over 43%, and resistance to broomrape and *Sclerotinia sclerotiorum* will be used to create commercial hybrids for ecological culture.

2. Background of organic farming

2.1. Definitions

The Ministry of Agriculture of Romania considered organic farming (similar to organic farming or biological agriculture), which differs fundamentally from conventional agriculture, as a "modern" process to cultivate plants, to fatten animals, and to produce food (www.mapam.ro) [5].

The Commission for Codex Alimentarius defines organic agriculture as "a production management system that promotes and maintains healthy development of agro-ecosystems, including biodiversity, biological cycles, and soil biological activity."

As science, organic farming deals with the systematic study of materials (living organisms and their environment) and functions (intra- and inter-relations material structures) of the agricultural systems, with design and management agro-ecosystems capable of providing for lengthy human needs for food, clothing, and housing, without reducing the potential environmental, economic, and social impact.

As occupation, organic farming is the activity that integrates theoretical knowledge about nature and agriculture in sustainable technological systems, based on the material, energy, and information resources of the agricultural systems (Toncea, 2000) [6].

To achieve this, organic farming relies on a number of objectives and principles, as well as on best practices designed to minimize human impact on the environment, while ensuring that the agricultural system operates as naturally as possible.

2.2. Principles underlying organic farming

Under the agreement in the integration of our country into the European Union, one of the measures imposed, inter alia, is the implementation of organic farming system. Apparently, this was something new, but some restrictions were easier to accept, for example, the interdiction for the use of chemical inputs that were not applied anyway on large surfaces in many agricultural areas due to economic considerations. However, a cause of concern is the lack of market demand for certified organic products and the low purchasing power of consumers. The price of an organic product is higher than its counterpart produced in the conventional system.

The normative acts operating in food production are particularly following the change in state of the art that occurred in agronomy. They do not refer solely on primary agricultural production sector, but also take into account the whole food chain, from primary production to final consumer. The agrifood complex is characterized by:

• Increasing the responsibility of those who practice this type of activity;

• Raising awareness and ability to reach market leadership.

Farms and organic agro companies are generally small- or medium-sized. Worldwide, most organic farms occupy small areas (0.5-30 ha), cultivate, and/or grow a small number of one, two, or three species of plants and animals and process one, two, or three different agricultural products.

Organic farming methods used in obtaining the unprocessed primary plant products, animals, and unprocessed animal products; animal and vegetable products processed for human consumption prepared from one or more ingredients of plant and/or animal origin; and compound feed and raw materials must meet the following conditions:

• Compliance with the principles of organic production;

• Non-use of fertilizers and soil improvers, substances used in animal nutrition, pesticides, food additives, growth promoters, cleaning and disinfecting products for livestock buildings, and products other than those permitted to be used in organic farming.

Developing of crop cultivation technologies targeted for alternative agriculture, especially for organic farming agriculture may improve the performance socio-economic indicators for these activities. This requires proper management of all the factors that contribute to high and stable yields per unit area, compliance with specific regulations and finally the recognition of finished products, in this case, an organic production certification.

2.3. Specific organic farming practices include

• Crop rotation as a prerequisite for the efficient use of farm resources;

• Very strict limits on chemical synthetic pesticides and chemical fertilizers, antibiotics for animals, food additives, and other substances used for additional processing of agricultural products;

• Not using of genetically modified organisms;

• Utilization of existing resources on site, such as using manure as fertilizer from animals and feed produced from the farm;

• Choice of species of plants and animals resistant to diseases and pests, adapted to local conditions;

• Livestock in freedom and open shelters and feeding them with organic feed;

• Using animal husbandry practices tailored to each race individually.

2.4. The objectives of organic farming

• Avoid all forms of pollution, both in products and in the environment;

• Maintain the natural fertility of soils, thereby ensuring food security in a sustainable planet;

• Allow farmers to have a decent life;

• To produce in sufficient quantities and at an appropriate quality level, thus ensuring the health of food consumers.

2.5. National and international legislations

The provisions on labeling of products from organic farming stipulated in Regulation (EC) no. 834/2007 on organic production and on labeling of organic products stated in Regulation (EC) no. 889/2008 that provide detailed rules for implementing Regulation (EC) no. 834/2007 are very precise and aim to offer consumers full confidence that products carrying the organic product label or the Community logo are obtained in accordance with the rules and principles contained in these regulations or, in the case of imports, are under the equivalent system with less demanding requirements.

To obtain and market labeled organic products and carrying specific organic production Community logo, producers must complete and strictly follow a rigorous process.

Thus, before you can obtain agricultural products that can be marketed as products of organic farming, the products must first undergo a conversion period of at least two years.

During the entire chain of production of an organic product, operators must constantly observe the rules established by Regulation (EEC) no. 834/2007.

In Romania, control and certification of organic products are currently provided by private inspection and certification bodies. They are approved by the Ministry of Agriculture and

Rural Development (MARD), based on the criteria of independence, impartiality, and competence as established in Order no. 688/2007 regarding the "Rules for organization of the inspection and certification system and approval of the certification and inspection bodies".

MARD's approval of control bodies requires a previous mandatory accreditation in accordance with European standard EN ISO 45011: 1998, which was issued by an agency authorized for this purpose. Following the inspections performed by regulatory bodies, certain products of operators complying with the rules of organic production may receive organic product certificate, and these products are permitted to be marked as "eco-labeled products".

Before the application of the label to an organic product, the following requirements must be fulfilled: the reference to organic production logo, name and code of the inspection and certification body that carried out the inspection and issued the organic product certification. The "ae" logo specific for national organic products, together with the Community logo, can be used for better views of consumer products from organic production.

The right to use the "ae" logo on product labels and packaging of organic products is given to producers, processors, and importers registered with MARD and holding a contract with a control body approved by MARD.

As part of the campaign to promote organic agriculture in the European Union (EU) at the initiative of the Directorate General for Agriculture and Rural Development of the European Commission, a website dedicated to this purpose was created: www.ec.europa.eu/agriculture/organic/home.ro.

The main objective of this site is to inform the general public about organic farming system as a starting point in the realization of promotional campaigns in different Member States.

Additionally, in order to promote the organic products, the European Commission provides support of up to 50% of information and promotion programs submitted by professional and inter-professional organisations, involving at least 20% of the actual cost of measures, and budget co-financing being provided by the State in accordance with Regulation (EC) no. 3/2008 on information and promotion actions for agricultural products on the internal market and in developing countries and Regulation (EC) no. 501/2008 that lays down detailed rules on implementing Regulation (EC) no. 3/2008 (information taken from the MARD website).

2.6. The national and international situations

If during the period 1950-1990 in Romania the objectives were to increase agricultural production to meet food requirements in view of the growing population, today the objectives are focused on finding new solutions that aim to respect the environment, create a system production that is economically viable, and maintenance and use of natural resources.

This new type of farming is called sustainable agriculture, and it involves a set of techniques and practices that should ensure a satisfactory production, ensuring food requirements are met and taking into account environmental protection.

After 1990, the gap recorded between quantitative indicators expressing the production potential and quality, caused by low endowment and equipment necessary to conduct the production process as well as related inputs, led to the development of technologies' extensive culture.

Another cause is the high fragmentation and dispersion of farms due to the implementation of the Land Law no. 18/1991. Currently, the farming land (14.8 million. Ha) is dispersed in about 40 million parcels. In 1972, the I.F.O.A.M. (International Federation of Organic Agriculture Movement) based in Germany was established. This federation groups more than 670 organizations and institutions from more than 100 countries worldwide.

The European Economic Community (EEC) recognized a majority vote of the European Parliament on 19 February 1986 on the existence of alternative agriculture based on resolutions adopted through Regulation 2092/91. A series of regulations were formulated, of which particularly important is Regulation EEC 1936/1995, which specified that from 1 January 2000, organic farming materials are the only ones to be used in sowing/planting.

According to I.F.O.A.M. statistics (February 2001), the world agricultural area intended for organic production was estimated to be 15.8 million hectares, with the largest area in Australia (7.6 million hectares), Argentina (3 million hectares), and Italy (1 million hectares).

In all EU countries, there is a real desire for developing OA, which will hold over 10% of the cultivated area. Agricultural area in the "bio" or "organic" agricultural systems in some countries is as follows: Italy - over 1.1 million ha, United Kingdom - 600,000 ha, France - 400,000 ha, Spain - 380,000 ha, and Austria - 250.000 ha. In the USA and Japan, about 20% of food is through organic production system.

In Romania, organically cultivated agricultural areas have seen a spectacular growth in the period 2010-2013, so at the end of 2013, about 301,148 ha were recorded by MARD.

Regarding the European organic food market, Germany has the biggest market, with sales of approximately 2.5 billion euro, and in terms of average consumption per capita of ecological products, Denmark and Switzerland are leading.

The markets for organic products are both the countries that depend on exports of organic products (Italy) and the countries that depend on imports of organic products (UK). Extremes of demand and supply in each country adjust by themselves. According to the study, the current situation appears to be changing because, in the UK, it is estimated that domestic production will meet the demand, while in Italy, the demand will increase. Today, increasingly more organic products are imported from Eastern Europe.

European Commission experts estimate that the market for organic products last year reached a value of 23 billion euro in the European Union. The organic market in the European Union is virtually all primary and processed agricultural produce (bread, wine, meat, milk, oil, fish, etc.). According to the study, organic products are generally 25-30% more expensive than conventional products, but depending on the supply and demand, the price could reach 400% of the price of the conventional ones.

Many local experts consider that countries in the Eastern Europe would need 10-15 years to be able to develop and structure the internal market at the level of the Western EU states. An argument invoked to support this assertion is the example of Spain, where it required about 17 years after integration to structure the internal market at the level of the other member states. Meanwhile, Spain exports almost all northern European market organic products. Eastern European countries will need to focus on organic production of the scanty products in the EU, including vegetable protein and red fruit, because Western countries have begun to significantly reduce production in sectors requiring a large labor force.

In Romania, the ecological production sectors benefit from European funding of about 200 million euro, which is available through a dedicated position in the new National Rural Development Programme (RDP) 2014-2020.

In addition, payments for OA, which are made by APIA, will continue. The registered farmers in organic agriculture will receive grants of 500 euro/hectare/year for growing vegetables, 620 euro/hectare/year for horticulture, 530 euro/hectare/year for vineyards, and 365 euro/hectare/year under organic cultivation of medicinal plants.

The experts appreciate that prices of organic products could be 10-20% higher than those of conventional ones if there are many farms and slaughterhouses certified. Romania currently has only 2-3 farms of laying hen organic certificates and some organic dairy farms, but instead the Romanian exports of organic wheat are significantly high, meanwhile part of this commodity is imported back as processed ecological products at prices 2-3 times higher than the conventional ones.

According to the MARD, the value of the domestic market of organic products in 2008 was about 20 million euro, while exports were at 100 million euro, which was twice the amount in 2006. Under an adjustment of Common Agricultural Policy (CAP) in 2009, Romania proposed that organic farming be financially supported by this package. Since this adjustment, CAP has created a financial reserve that allows the Member States to develop certain programs to fully support a particular context, technically called Article 68.

The financial envelope allocated to Romania for 2010 only amounted to EUR 5 million. The increase in the organic market in Romania continues; with only 86 registered organic food processors in 2008; in 2010, the number became 3,155; and in 2012, it was 15,194.

Exports of organic products in 2008 amounted to 100 million euros, which was equivalent to about 130,000 tons of products, of which only 1% were processed products and 0.94% were honey products. The primary export destinations were the Netherlands, Germany, Denmark, Italy, and the UK. Imports of organic products were worth 10.8 million euro, which was almost double the amount for 2007, with fruit and legume preserves, coffee, and sweets being the most significant products.

The turnover in organic agriculture worldwide was 46 billion dollars in 2007, up by 10% compared to 2006, while in Europe the figure reached a level of 15.4 billion euro, 15% more than in 2006.

The productive potential of agriculture ecological system of the country can reach up to 15-20% of the total agricultural areas largely concentrated in hilly mountain where technology maintenance and use of pastures were based on traditional methods - organic (manure application, utilization of grazing and/or mowing, use of fodder and clover ameliorating soil fertility, use of vegetable-livestock mixed system), but are not neglected arable land in the North-East.

At global level, two opposite trends are rising as an increasing concern:

a. **Overproduction** and negative side effects of industrial type of farming that include decreasing of soil fertility due to erosion, acidification, salinization, and exhaustion of the reserve of organic matter; reducing of biological and genetic diversity; increased risk of air pollution exhaust and ammonia, shallow and deep waters and soils with nitrates, and heavy metal contamination of food with toxic substances, etc.;

b. **Production for subsistence** and its negative consequences - hunger and social inequity.

These imperatives can be resolved only by organic farming, an agricultural practice in some countries that is called organic or biological farming, which sprang from the secular experience of agriculture.

Organic farming is not a miracle or a wonder, but a creation of nature-loving farmers, who aim for harmony and dynamic interactions among soil, plants, animals, and humans, or, in other words between supply natural ecosystems and human needs of food, clothing, and housing.

2.7. Practical aspects of OA

- The agro-ecological systems have long life due to components, structural and functional stability, and ability to cope with any disruptive or disturbing factor.

- Organic production is done on farms, individual households, family associations, agribusiness companies, and rarely in large agricultural associations and companies or holding. Organic products are obtained also in the aquatic environment, forest, and other natural systems.

- Generally, many agricultural and agro-ecological farms are in small- or medium-sized category. At world level, the average surfaces for organic farms are within 0.5 and 3.0 ha range, and most of them cultivate only 1-3 different agricultural species.

- All organic farms and agro-industrial societies undergo a longer or shorter conversion period, which is equal to the time between the start of ecological management and getting the certificate by the ecological farm or company.

- Certification is provided by a national or international organization that is recognized by the International Accreditation Service International Federation of Organic Agriculture Movements (IFOAM) and empowered to assess and guarantee in writing that its production or processing system is in compliance with the standards of organic agriculture.

- The transition from conventional to organic farming is done step by step, in order to protect the economy from the shocks of decreases in productivity, and to allow producers to gain confidence in the ecological systems. Certification of these business units is made as soon as a part of their work meets the environmental standards and provided that the two systems (conventional and organic) are clearly separated both in documentation and in production.

- With very few exceptions, organic farms are mixed, plant-animal type, on the one hand, to capitalize on higher crop and, on the other hand, to reuse as much of the nutrients extracted from the soil by plants grown. In this case, the structures of animal species and categories are determined by the potential of the farm and vegetable farming area, as well as the economic and financial resources (buildings and plant breeding, money) and the manpower (number of people, age, training) available in the farm.

Exceptions to this rule are organic vegetable farms and processing and marketing firms for semi-organic products. In such cases, the bulk of production is for direct human consumption (vegetables, fruits, canned vegetables and meats, cheeses, vegetables, and animal extracts); processing of the products is done with minimum consumption of energy, and this energy is, as far as possible renewable, sourced from animal manure (biogas), wind, and local fluid (residues and organic waste).

- The activities of farms and agro-industrial companies are carried out according to international and national rules. Any deviation from these standards results in losses, including the loss of farm and the ecological society.

- In organic farms and processing companies, all species and varieties of domesticated plants and animals are grown and processed, except those created by genetic engineering.

- Farms and processing companies of organic product are using mostly own financial economics and social resources. Land, goods, and services of the agro-ecological units are mainly privately owned, and the funds are secured, for the most part, from its own resources. In countries with developed organic farming, a significant share of financial resources is provided by the state through a diversified mechanism of subsidies (exemption from taxes, production inputs, and additional expenses to conventional agriculture). The workforce consists of organic agro-industrial unit farmers or owners and close relatives.

- Some farms and agro-industrial companies undertake labor from outside, also, but only for a determined period of time when the workload exceeds the skill and strength of the permanent employees.

Regarding the problems of agro-ecological systems, Köpke (2005) argued that compared with intensive farming system, ecological system is characterized by:

- Reduced availability of nutrients, especially nitrogen and phosphorus, with consequences on the level of yields (because of the limited growth) and especially on their quality;

- The danger of a high level of weed and pest infestations due to absence of chemical treatments. Claude Aubert, one of the pioneers of organic farming, supports with scientific arguments that for organic farming "the genotype is more important than the whole technology".

2.8. Reference of knowledge on the topic addressed

The Intergovernmental Panel on Climate Change (IPCC), which brings together experts from around the world, published on 6 April 2007 in Brussels a new report on the impact of global warming on people and the earth. This report is a readjustment of the report in 2001 and is recognized by 192 UN member states. The crucial passage of the new report indicates that "a drastic change in climate is expected if carbon concentrations in the atmosphere will reach 550 ppm (parts per million), which would cause a rise in temperature of about 3 degree Celsius. The main consequences of global warming are increasing ocean levels and extreme weather events (heat waves, droughts, floods, strong winds) that will bring major impacts like disappearance of animal and plant species, increasing human health risk, and inevitable demographic changes. Crop yields fluctuate from year to year, and this is being significantly influenced by climate variability and extreme weather events. Climate variability impacts all sectors of the economy, but agriculture remains the most vulnerable.

In Romania, from about 14.7 million ha of agricultural land, of which 9.4 million ha is arable land (64% of arable land), 7 million ha of agricultural surface (48%) soils are affected in different degrees by frequent droughts in most of the years and more than 6 million ha of agricultural land are affected by excess moisture in wet years. The extent and intensity of extreme weather events decrease annual agricultural production by at least 30-50%, and sustainable conservation of natural resources in agriculture is necessary to ensure scientific validity of all actions and measures to prevent and mitigate the consequences. Drought is a natural phenomenon caused by insufficient rainfall for meeting the crop requirements. The impact of drought is influenced by the severity of drought, physiological status of crop (including the development stage and cultivar adaptation) and soil properties.

The most severe effects are manifested especially on the rural population dependent on farming. Global climate changes as manifested by the increasing average temperature and change in rainfall regime have led, in recent decades, to an increase in drought-affected areas worldwide. In Romania, the areas most vulnerable to extreme drought are the south-eastern and Dobrogea, Baragan and southern areas of Oltenia, Muntenia, and Moldavia.

The term "desertification" refers to reduction or destruction of the biological potential of land that can lead to problems similar conditions in desert areas. Desertification includes the interaction of large-scale global climate dynamics, reflecting the general circulation of the atmosphere and ocean and climate physics of the earth's surface. It can be a result of the interaction of natural recurrence of droughty years with practice of irrational exploitation of the land, deforestation, and intensive grazing. Climatic data from the past century show a gradual warming of the atmosphere and a significant reduction in rainfall as limiting factors for crop growth and productivity and utilization of water resources. These changes can have significant impacts on growth and development of crops during the growing season, depending on the intensity of the disruptive factor, the manner and duration of action, and plant species vulnerability to extreme weather events during production.

Globally, according to studies, a significant warming in the coming decades is expected as a result of increased CO_2 concentration in the atmosphere and significant changes in precipitation. The IPCC report (2001) estimated an increase in global average temperature from 1.4°C to 5.8°C by 2100, depending on the emission scenario, which is 2-10 times more pronounced compared to the current condition. The amount of rainfall will record a rise/fall trend of between 5% and 20% globally, with significant differences occurring especially at the regional level. It will also intensify the occurrence of extreme weather conditions (winter and summer extreme temperatures, droughts, floods, tornadoes, hurricanes, etc.) with major consequences on the entire planetary ecosystem.

In Romania, projections of global scenarios for the period 1991-2099 as compared to the period 1961-1990 revealed an increase in the average air temperature of about 2°C during winter and 3.5°C to 4.3°C during summer (3.5°C and 4.3°C in the north and south, respectively). With regard to precipitation the expected changes are insignificant during summer and winter will be recorded water deficits. The northwest country regions are expected to become slightly wetter meanwhile the southwest and central regions will become drier.

In the twentieth century, global warming shows an annual average temperature rise of 0.3°C in almost the entire country, with the increase in temperature being more pronounced in the southern and eastern areas. Significant warming was experienced during winter and summer seasons (with Bucharest-Filaret being the most pronounced, 1.9°C), and significant cooling was found during fall in the western regions of the country.

Regarding the distribution of precipitation within year, there was a downward trend in the annual quantities especially in the central regions, and during the winter season, a decreased precipitation was observed in most regions, being more pronounced in the south and west.

Effects of global warming further include the following changes in the occurrence of meteorological phenomena in hot or cold season of the year: increased frequency of tropical days, decrease in the frequency of winter days, increasing average maximum temperature during winter and summer (up to 2.0°C in the south and southeast), significantly decreased thickness of snow in the Northeast and West, and increased annual production of winter atmospheric phenomena (frost, ice, frost).

Today, global climate change is associated with increased pollution, deforestation or changes in the landscape that caused an amplification on the process of aridization. As a result, some high-risk areas for drought tend to be affected by aridity and even by desertification (disappearance of vegetation cover and soil degradation). In our country, the high-risk territories for drought, with a tendency to be affected by aridity and desertification, include large areas of Dobrogea and southern Romanian Plain. These areas may be classified as areas most vulnerable to excessive and prolonged drought.

In the next decades, the implications of global warming in the industrial economy, water supplies, agriculture, and biodiversity will be very obvious. Globally, therefore, it has the effect of warming and increased frequency and intensity of extreme events, especially droughts and floods. The causes that lead to these phenomena are evident about both climate and human interventions or wasteful use of land and water resources, inappropriate agricultural practices, deforestation, overgrazing, and air and soil pollution.

During extreme droughts, the current agricultural practices recommended are: fixing assortment of varieties and hybrids at the beginning of each crop year and the use of appropriate technology depending on soil water reserves from sowing; cultivating a greater number of varieties/genotypes with different growing season for better use of the climate conditions, especially moisture regime. Significant yield losses can be prevented through observance of recommended sowing period, irrigation or application of a minimum tillage system, utilization of varieties adapted phenologically to the new climatic conditions (in order to avoid the occurrence of critical phases as pollination and grain filling during the maximal stress periods) and better adapted physiologically to stress.

In the long term, the necessary measures for the prevention and mitigation of climate change include reforestation programs, reducing pollution, restoring and upgrading anti-erosion work, and expansion of the development and improvement of sandy soils, etc. At the same time, educating people and raising awareness on environmental protection are major requirements in developing adaptation strategies to climate change.

Solutions and recommendations for the development of actions and procedures to prevent and minimize the effects of climate variability in agriculture must include the already well-known whole complex of measures (agro-technical, cultural, irrigation, etc.) and carrying out swift action and intervention to limit the consequences and spatial extension of the affected area.

However, addressing issues related to climate change impacts requires specialized scientific data and analysis, risk management in agriculture mainly involving actions concerning the management and conservation of environmental resources, and making the right decisions in the right perspective.

In 1996, the National Commission on Climate Change/CNSC (HG 1275-1296) was established, and in July 2005, Romania's National Strategy on Climate Change (GD 645/2005) was approved. Also, with Law no. 111/1998, Romania joined the United Nations Convention to Combat Desertification (CCD); the Convention adopted in Paris on 17 June 1994 the declaration that 17 June be recognized as the Desertification and Drought Day worldwide. (SMART financial - www. SMARTfinancial.ro.[7])

3. Improving sunflower to biotic and abiotic stress factors

NARDI-Fundulea has obtained an invaluable genetic basis with over 50 years of research experience. NARDI-Fundulea is the basic institution in Romania that provides the necessary seeds and parental lines of traditional culture system.

The requirements for sourcing germplasm to improve sunflower hybrids are becoming bigger and more important. The greater diversity is conserved, the more chances to meet the present and the future. Loss of genetic diversity or genetic erosion may occur as a result of many interrelated causes, such as socio-economic and agricultural, natural disasters such as epidemics, long periods of drought and floods, and even human contribution.

The collection, evaluation, and preservation of wild species of sunflower were done carefully and were the basic objectives of the scientific cooperation of the FAO Network Research Sunflower Sun. From its foundation in 1975 until today and in Vrânceanu's (2000) study [1], the main objective of improving the sunflower hybrid is said to be further improving productivity by increasing seed production and seed oil content. After seed production and oil content, the major objective of improvement is: genetic resistance to disease *(Sclerotinia sclerotiorum, Phomopsis helianthi)*, parasite *Orobanche sp*, drought, and heat; with less attractively for birds were obtained.

3.1. Material and methods

The genetic materials used were seven inbred lines of sunflower obtained at the NARDI-Fundulea and wild *H. argophyllus* known as resistant to drought.

Two locations were chosen for organic testing: Stupina (Constanta), known as pole drought in Romania, and Fundulea (Calarasi county).

The breeding methods used were: interspecific hybridization (first year of experimentation), embryoculture to save interspecific *embryo rescue* backcross, self-pollination, and selection. Two generations/year worked in the field and in the greenhouse, as illustrated below.

As we have a lot of data from all the years of experimentation, we will present only the results obtained in 2014, which was an extremely dry year in terms of ecological culture of the Stupina. Some of the results were published in international journals, while others are in print.

Regarding drought tolerance, we deduced the parameters of productivity (weight/head, TKW, and oil content), which will be presented for each new genotype obtained.

Each "slash" code (for example 1/1/1...n) inside graphs represents a descendent from an initial interspecific hybrid that further was subject of the general breeding scheme (individual selection, self-pollination, back cross and new selection scheme). The labels Stupina1-Stupina3 and Fundulea1-Fundulea3 represent the number of repetitions per location.

3.2. Results

From Figure 1, we can see great differences among some genotypes, even if they have the same lineage. In the same cross-breeding, it has been observed that every head is a distinct genetic entity. Therefore, the seed obtained from each phenotypically different head was seeded (three times/head) into one isolate plant/row.

Both lines (1/1/1 and 1/1/2) showed instability of seeds and head weights, both at Fundulea and at Stupina. For TKV, 1/1/1 in terms of Fundulea, showed some stability; the differences between the three plants are insignificant. For Stupina, although TKW values are reduced by approximately 50%, stable new lines show the three plants. In Figure 1, the same applies. Great unevenness and instability on seed production/head were observed. The fact that while seed production was low, except for plant Fundulea 1 (1/2/1), TKW recorded values were between 29 grams and 49 grams.

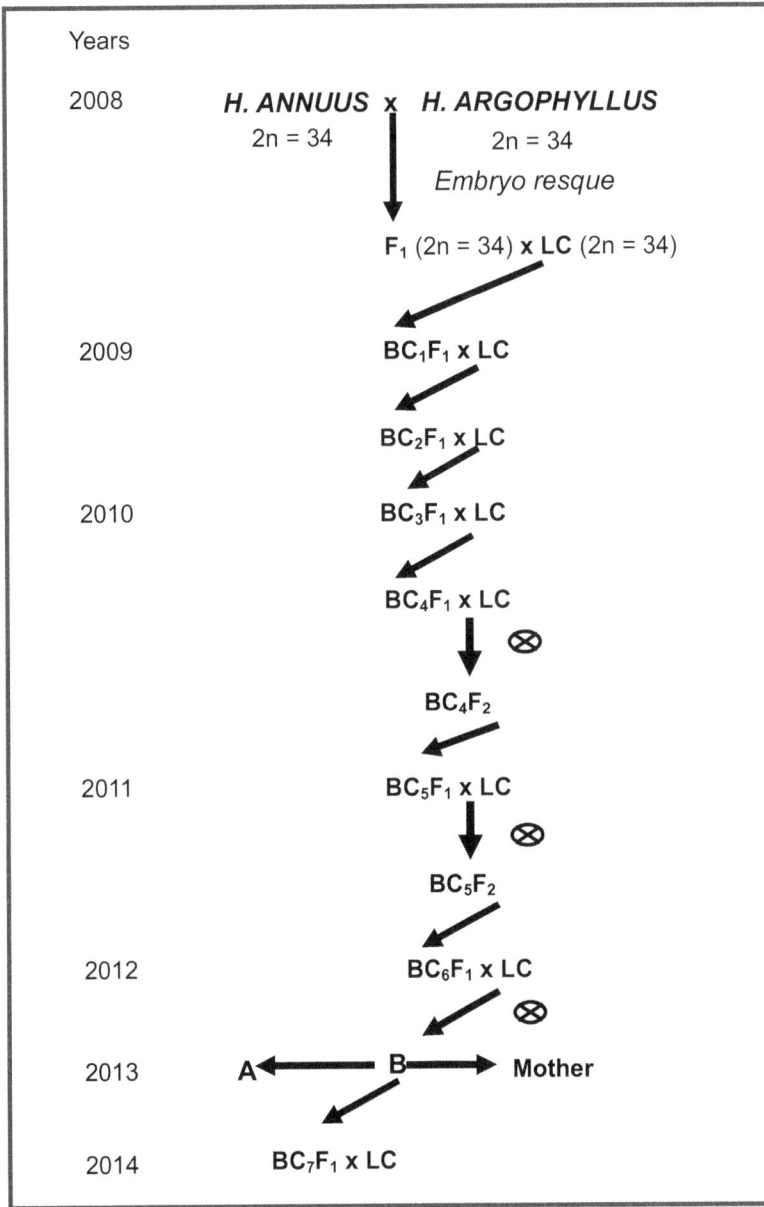

Figure 1. Sunflower breeding scheme

From Figure 4, one can see that the plant "Fundulea 3 (1/3/1)" progenies with a head weight of 70 grams at Fundulea and 38 grams at Stupina. The TKW for this genotype was the highest in both locations (49 grams).

Due to the very low values for both characters, in both locations, the genotypes 1/4/1 and 1/4/2 were not considered for the process of breeding for commercial hybrids (Figure 5).

In the extremely droughty conditions from Stupina, even the TKW and head weights were lower than in Fundulea they displayed a better uniformity.

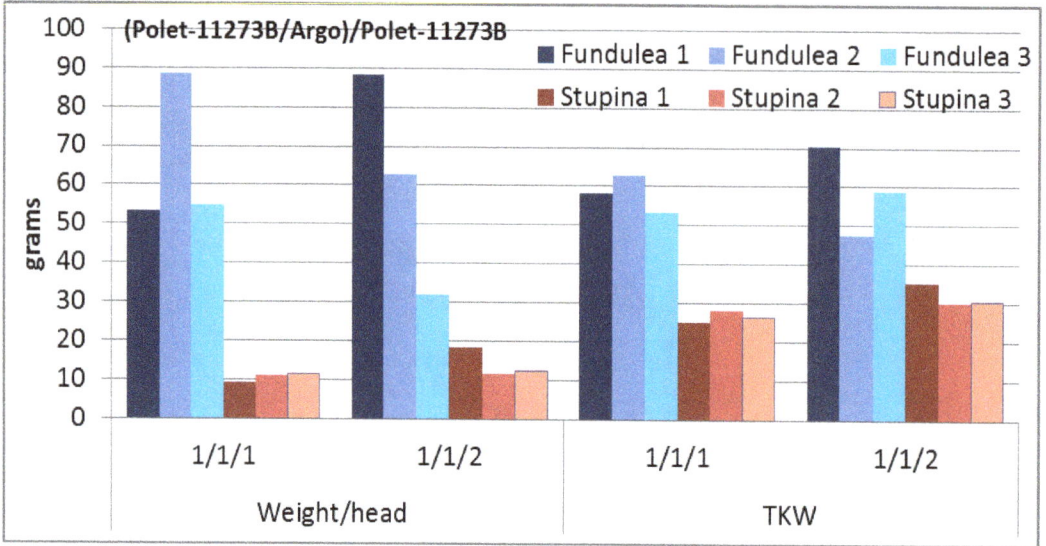

Figure 2. Average weight of head and TKW for the progenies of the progenies from backcross 7th generation of the 1/1/1 and 1/1/2 lines resulted from interspecific hybridisation Polet-11273B x *Helianthus argophyllus*

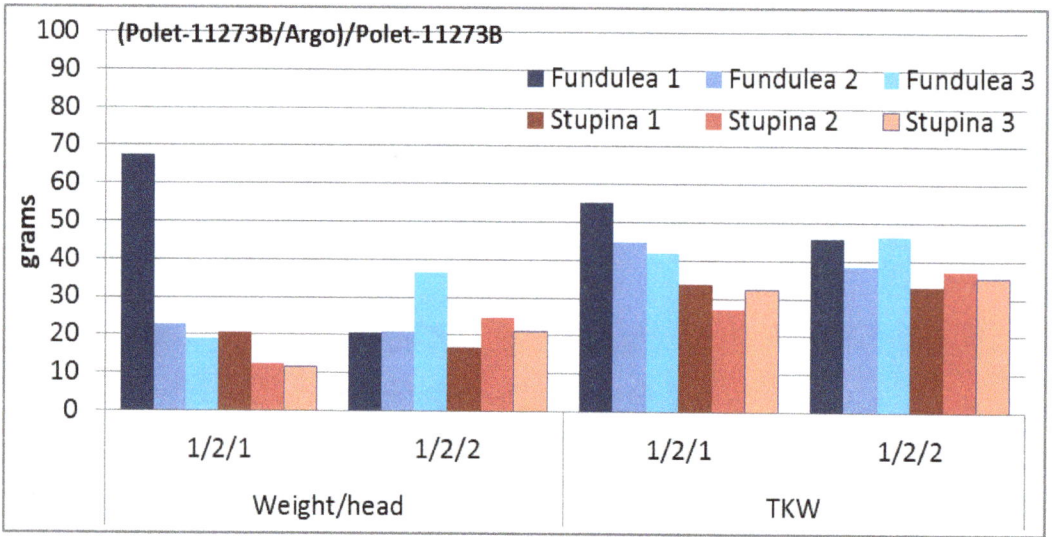

Figure 3. Average weight of head and TKW for the progenies of the progenies from backcross 7th generation of the 1/2/1 and 1/2/2 lines resulted from interspecific hybridisation Polet-11273B x *Helianthus argophyllus*

Due to the fact that the oil percentage is higher under heat and drought stress conditions, it is not surprising that all the genotypes obtained from hybridization of Polet-11273B x *Helianthus argophyllus* (Figure 6) have an oil content (estimated by NMR) greater than 40% at Stupina, significantly exceeding the oil content (determined with the same method) of the seeds obtained at Fundulea (29-33%)

From the hybridisation of line O-7493B with *Argophyllus*, 3 descendants with yield and oil content stability were selected: 3/1/1 (Figure 6); 3/2/1/ (Figure 7); and 3/4/2 (Figure 8). The seed

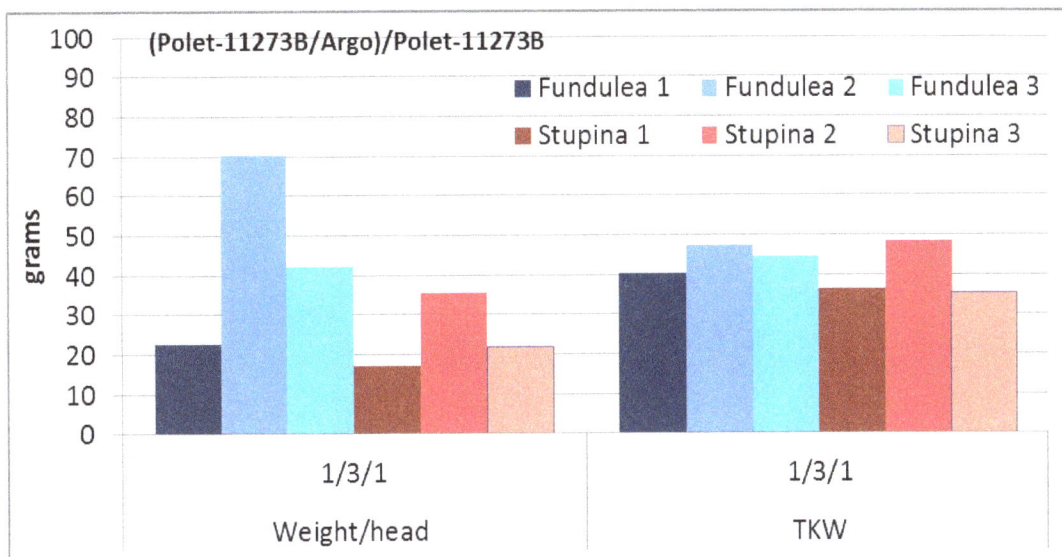

Figure 4. Average weight of head and TKW for the progenies of the progenies from backcross 7th generation of the 1/3/1 and 1/3/2 lines resulted from interspecific hybridisation Polet-11273B x *Helianthus argophyllus*

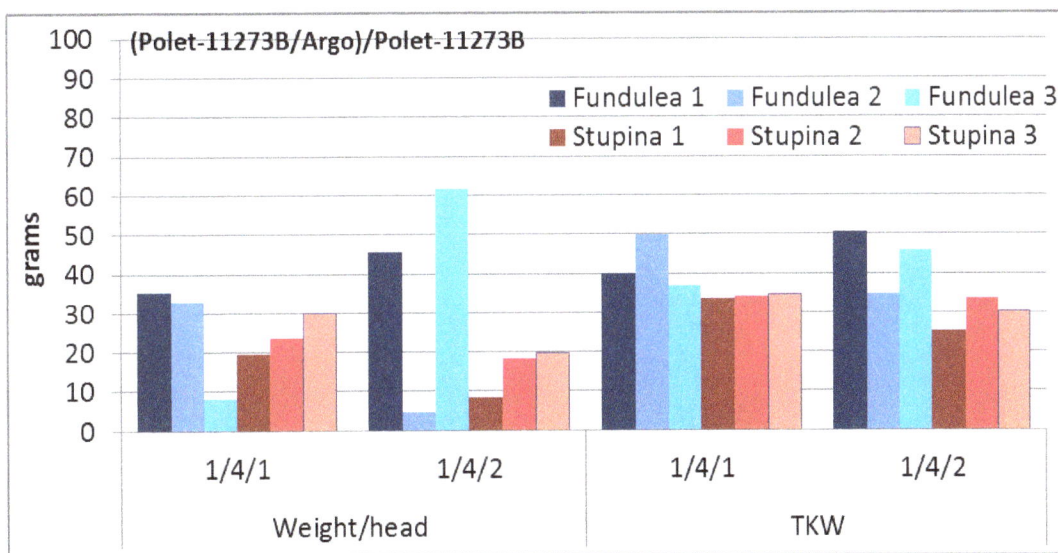

Figure 5. Average weight of head and TKW for the progenies of the progenies from backcross 7th generation of the 1/3/1 and 1/3/2 lines resulted from interspecific hybridisation Polet-11273B x *Helianthus argophyllus*

weight of heads of these descendants varied between 73 and 120 grams/head at Fundulea and between 25 grams and 60 grams at Stupina. The oil content varied between 32% and 40% at Fundulea and between 39% and 44% at Stupina (Figure 9).

After hybridizations, self-pollination, backcrossing, and selection of descendants of the hybrid Tard/85-19982B X *Helianthus argophyllus*, we obtained 11 new lines with a very large variability for the studied characters.

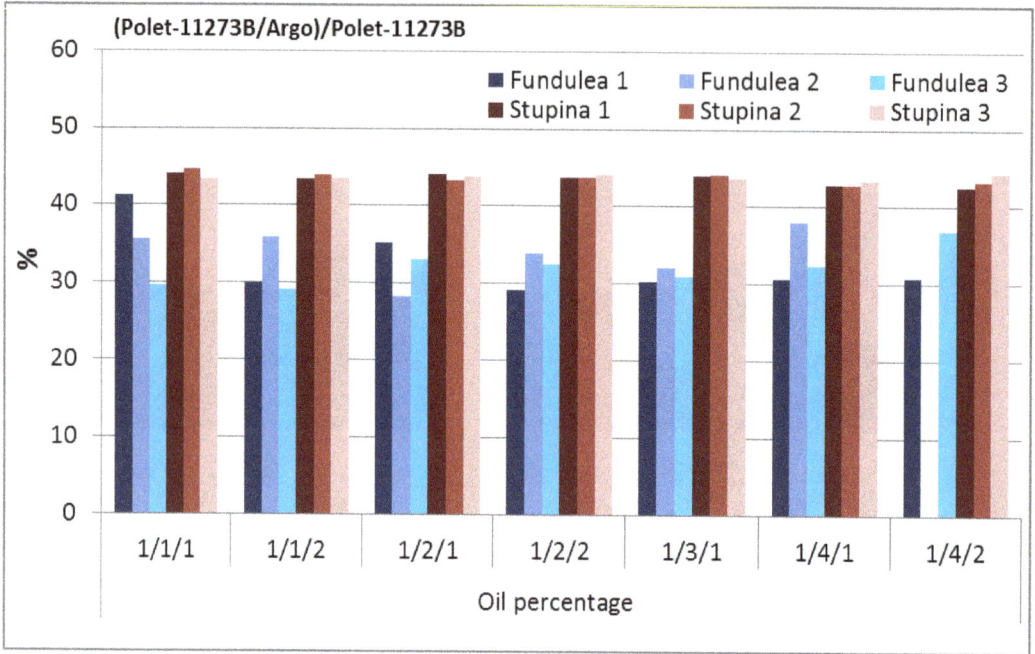

Figure 6. Oil content of the backcross 7th generation of sunflower lines obtained from interspecific hybridisation between Polet-11273B and *Helianthus argophyllus*

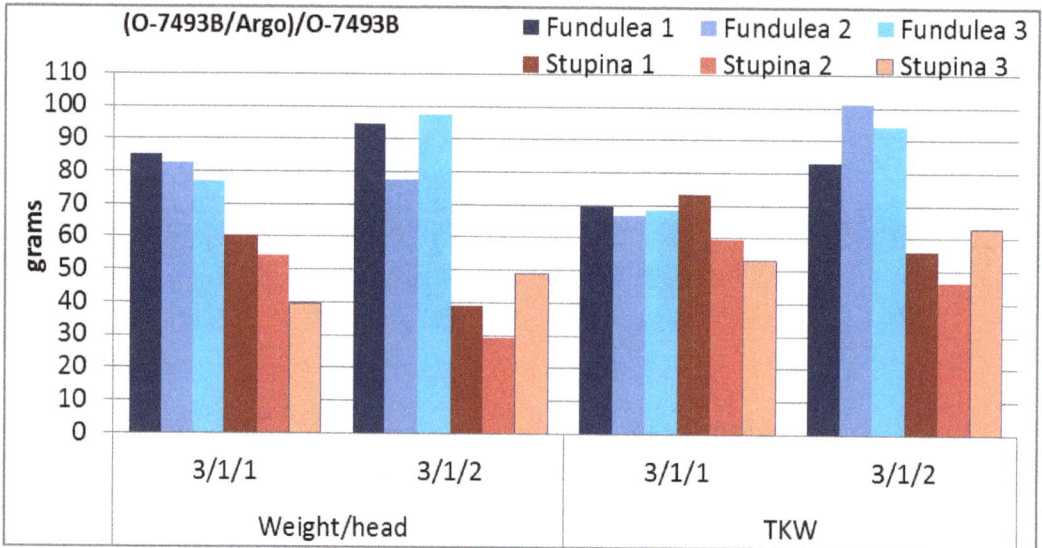

Figure 7. Average weight of head and TKW for the progenies of the progenies from backcross 7th generation of the 3/1/1 and 3/1/2 lines resulted from interspecific hybridisation between O-7493B and *Helianthus argophyllus*

Figure 11 shows that the weight of seeds/head in the case of line 11/1/1 was higher in the drought condition of Stupina. Therefore, the genotype "plant Stupina 2" achieved 72 g/head compared with "Fundulea 2" that produced only 40 g/head. Additionally, for the TKW character, this line proved to possess good adaptability to drought, reaching or exceeding the

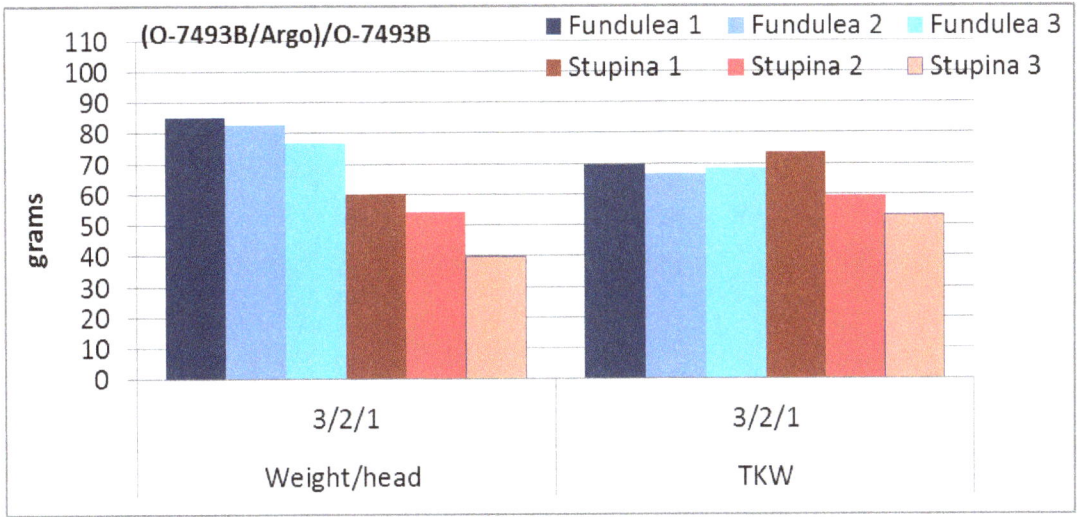

Figure 8. Average weight of head and TKW for the progenies of the progenies from backcross[th] generation of the 3/2/1 line resulted from interspecific hybridisation between O-7493B and *Helianthus argophyllus*

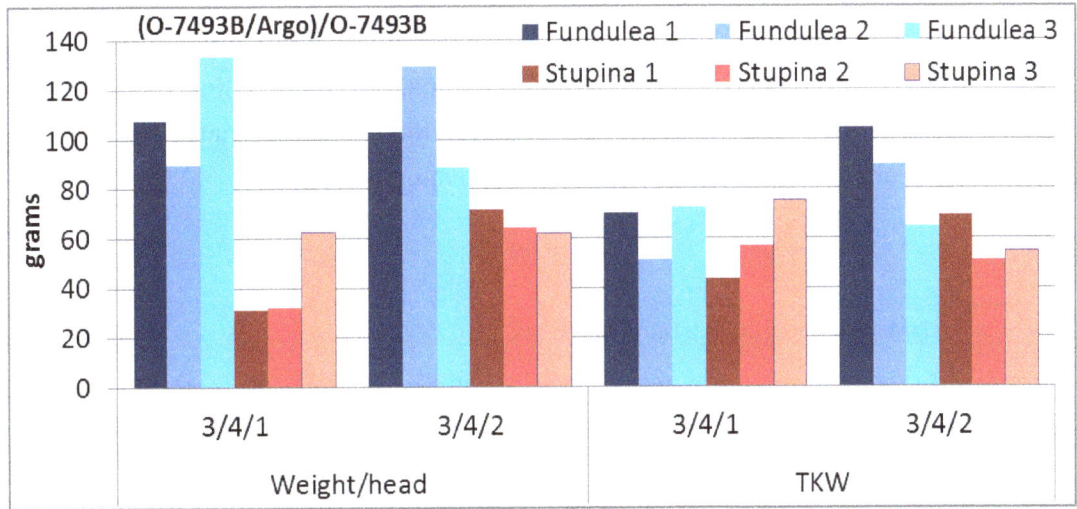

Figure 9. Average weight of head and TKW for the progenies of the progenies from backcross 7th generation of the 3/4/1 and 3/4/2 lines resulted from interspecific hybridisation between O-7493B and *Helianthus argophyllus*

values obtained at Fundulea, where the weather conditions were closer to normal. Another descendent of this interspecific hybridization is the line 11/2/1 (Figure 12) that achieved through the genotype "plant Stupina 2" a seed yield per head of 90 g and a TKW of 59 g being the only line out of all the combinations that have proven under drought conditions such a performance. It is necessary to mention that the line Tard/85-19982B is known to be like an intensive line with high yield under good irrigation and fertilization. In this case, it is obvious that the resistance and adaptability to drought were transmitted from the wild species, due to the fact that agro-ecological selection field from Stupina was not irrigated, and no fertilizer was applied.

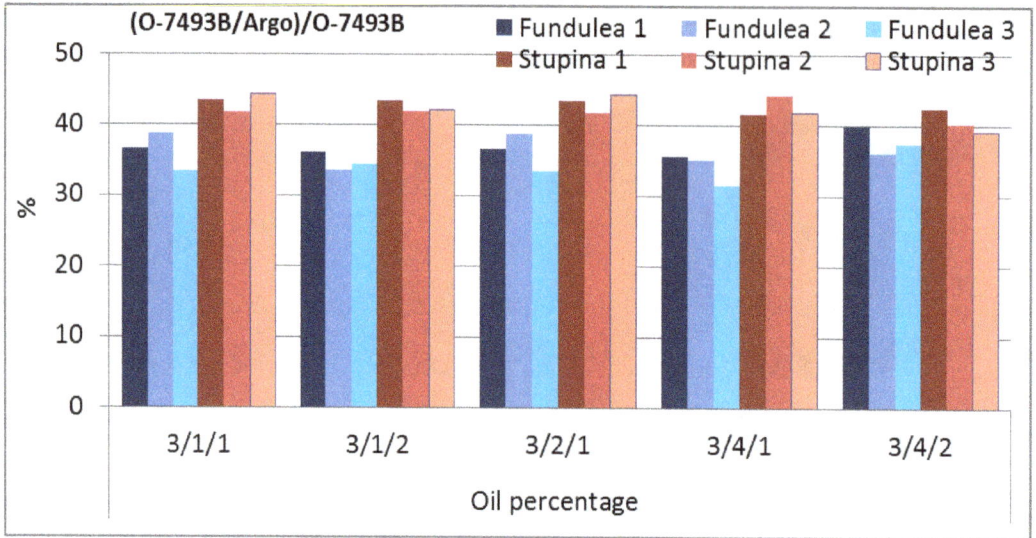

Figure 10. Oil content of the backcross 7th generation of sunflower lines obtained from interspecific hybridisation between O-7493B and *Helianthus argophyllus*

These two lines originating from this combination will be used to obtain commercial sunflower hybrids. For all other lines resulting from this combination, the breeding process will be continued through self-pollination and backcrossing, due to the fact that they represent a valuable biological material that can be further improved.

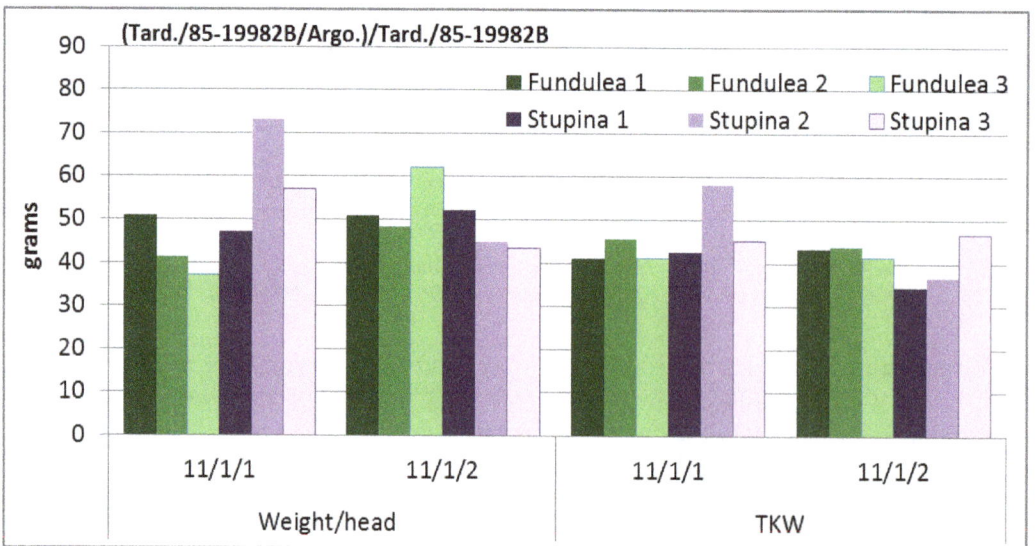

Figure 11. Average weight of head and TKW for the progenies of the progenies from backcross 7th generation of the 11/1/1 and 11/1/2 lines resulted from interspecific hybridisation between Tard./ 85-19982B and *Helianthus argophyllus*

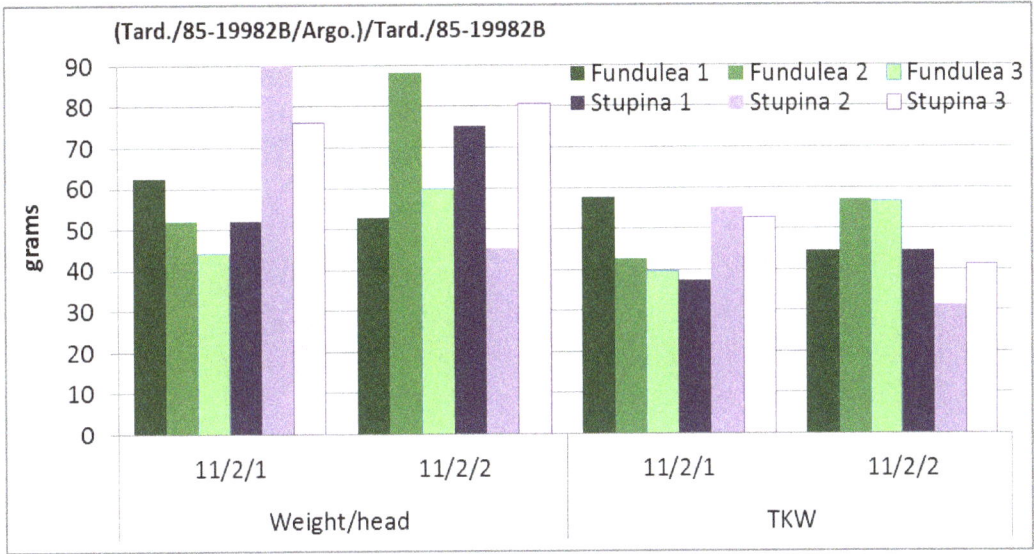

Figure 12. Average weight of head and TKW for the progenies of the progenies from backcross 7th generation of the 11/2/1 and 11/2/2 lines resulted from interspecific hybridisation between Tard./85-19982B and *Helianthus argophyllus*

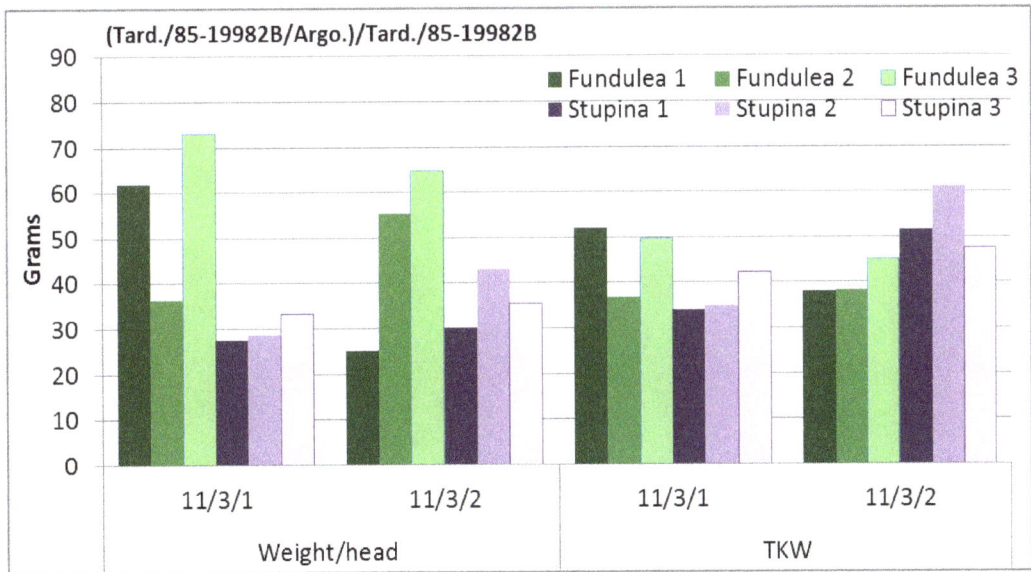

Figure 13. Average weight of head and TKW for the progenies of the progenies from backcross 7th generation of the 11/3/1 and 11/3/2 lines resulted from interspecific hybridisation between Tard./85-19982B and *Helianthus argophyllus*

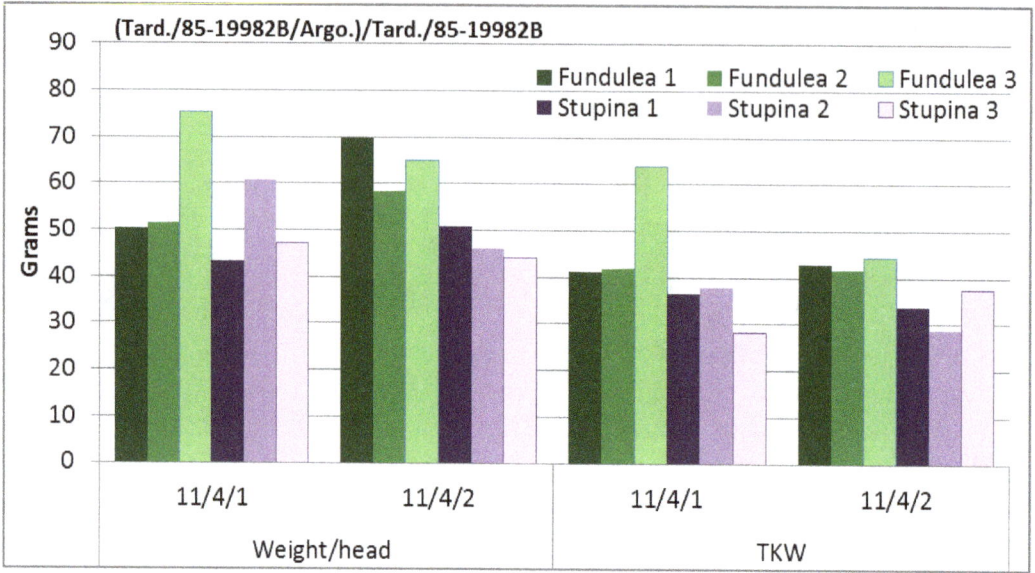

Figure 14. Average weight of head and TKW for the progenies of the progenies from backcross 7th generation of the 11/4/1 and 11/4/2 lines resulted from interspecific hybridisation between Tard./85-19982B and *Helianthus argophyllus*

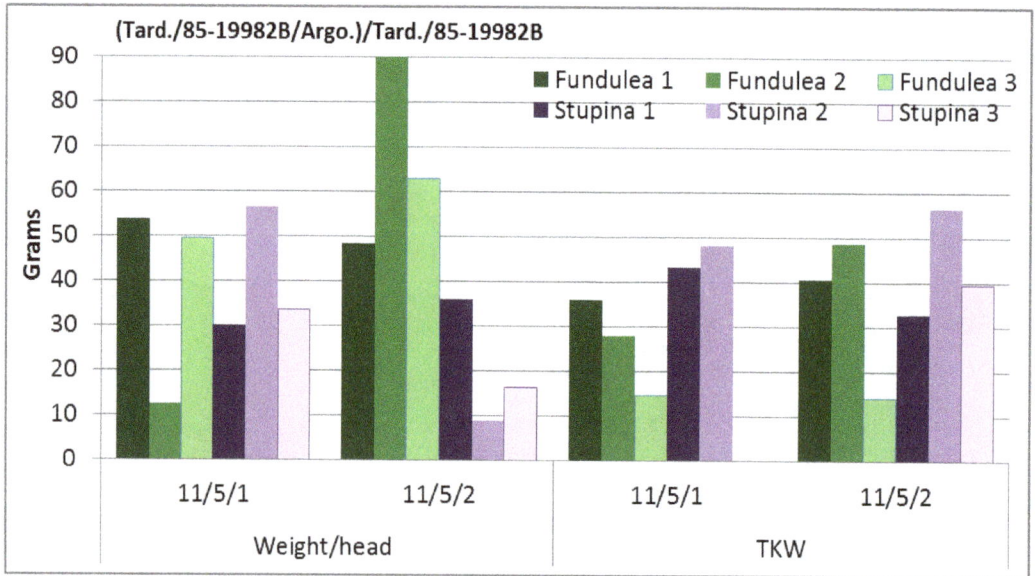

Figure 15. Average weight of head and TKW for the progenies of the progenies from backcross 7th generation of the 11/5/1 and 11/5/2 lines resulted from interspecific hybridisation between Tard./85-19982B and *Helianthus argophyllus*

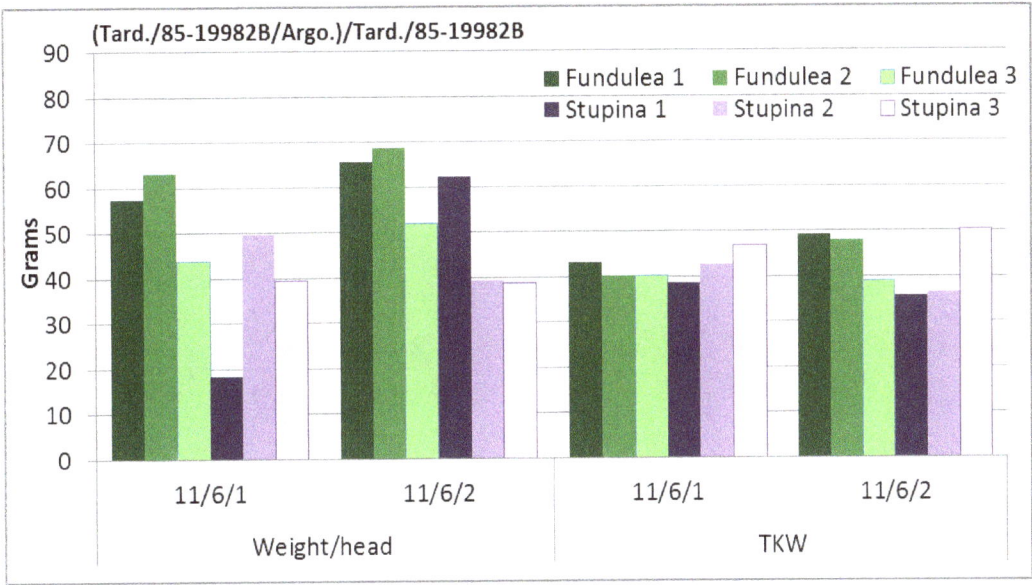

Figure 16. Average weight of head and TKW for the progenies of the progenies from backcross 7th generation of the 11/6/1 and 11/6/2 lines resulted from interspecific hybridisation between Tard./85-19982B and *Helianthus argophyllus*

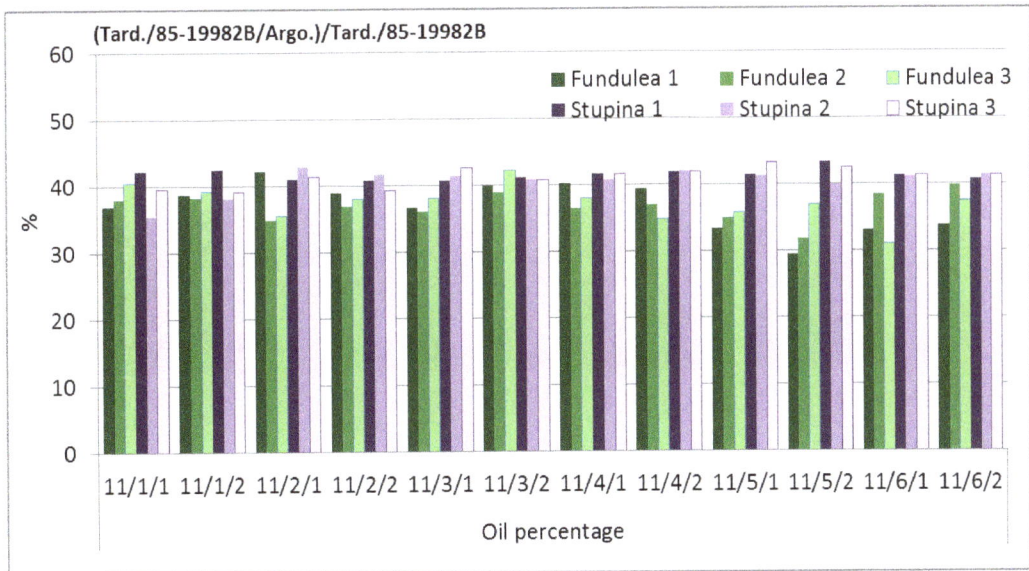

Figure 17. Oil content of the backcross 7th generation of sunflower lines obtained from interspecific hybridisation between Tard./85-19982B and *Helianthus argophyllus*

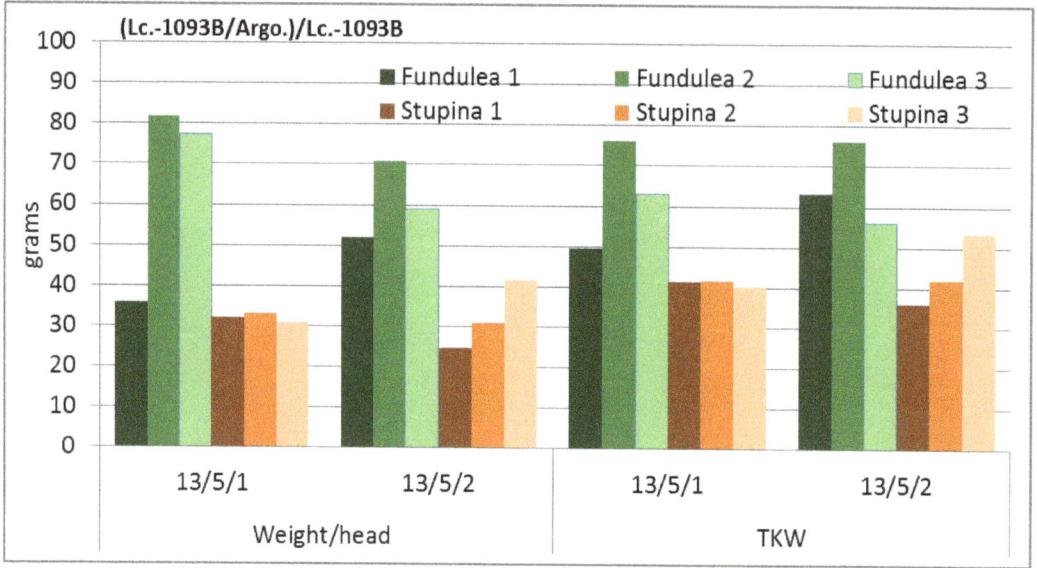

Figure 18. Average weight of head and TKW for the progenies of the progenies from backcross 7th generation of the 13/5/1 and 13/5/2 lines resulted from interspecific hybridisation between LC-1093 B and *Helianthus argophyllus*

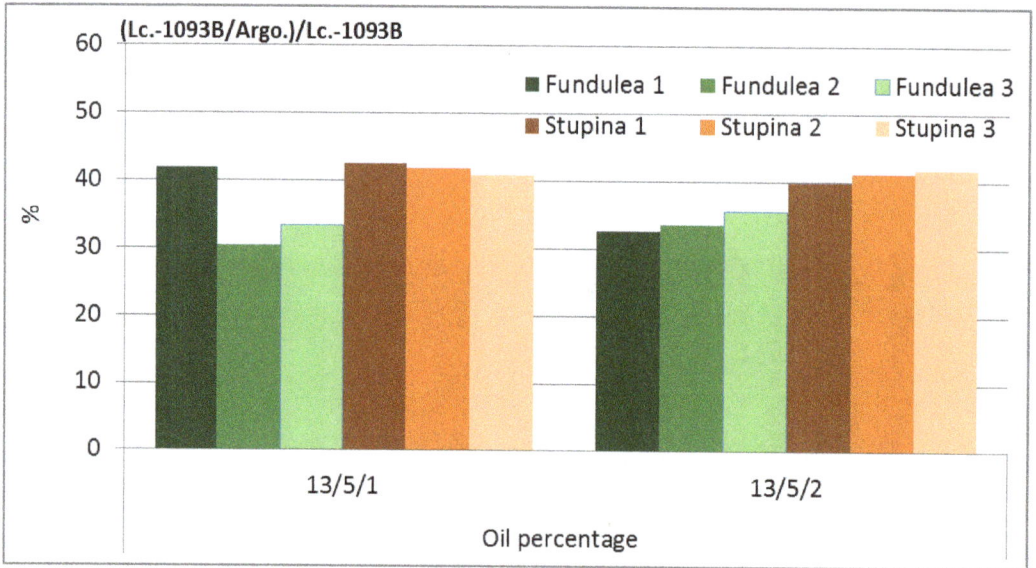

Figure 19. Oil content of the backcross 7th generation of sunflower lines obtained from interspecific hybridisation between LC-1093 B and *Helianthus argophyllus*

The inbred line 1093 B was considered by breeders as having a large ecological plasticity, and it is used in obtaining very valuable hybrids with resistance to plant diseases and *Orobanche*. In combination with *Argophyllus*, the results were spectacular. Even if the yield was very low under limited water conditions (Figure 17), the lines 13/5/1 and 13/5/2 proved a very good resistance to *Sclerotinia sclerotiorum* (at Fundulea) and bird atttack (at Stupina). It is necessary that at Stupina and under the agro-ecological management, there are enough tree windscreens were a lot of rooks and house sparrows are nesting and increasing very much their numbers. For the local farmers, these birds are source of damages not only for sunflowers but also for wheat and barley. This new line has the advantage that it is avoided by birds so even if the yield is low it is safe.

4. Conclusions

a. It is very important that together with the fulfillment of the main objective of the study (yield stability in water stress conditions in organic farming system), the results selection included achievements for biotic factors (resistance to *Sclerotinia* and *Orobanche*). Three genotypes with resistance to *Orobanche* in conditions of soil were identified, with a very high infestation with broomrape due to monoculture of sunflower for three years.

Under our experimental conditions, the genotypes with a longer vegetation period presented a better resistance to broomrape (Figures 20-21).

b. Some of the genotypes resulting from the interspecific hybridizations with *H. argophyllus* were not affect by the massive bird attacks from Stupina in 2014, when many farmers reported severe losses due to birds. This represents an important step in releasing sunflower hybrids with resistance or tolerance to this character (Figures 22-23).

c. In the conditions from Fundulea, in favorable year, a strong attack of *Sclerotinia sclerotiorum* was recorded, and this was a good opportunity to find among the tested combinations the reactions ranging from being tolerant to being sensible to this pathogen (Figures 24-25).

Figure 20. O-7493B X*Helianthus argophyllus* - genotype with sensitivity to *Orobanche cumana* and vegetation period of 115 days (Stupina location)

Figure 21. LC-1093 B X*Helianthuus argophyllus* - genotype resistant to *Orobanche Cumana* and vegetation period of 130 days (Stupina)

Figure 22. Tard/85 -19982B - genotype with seeds highly preferred by birds (Stupina)

Figure 23. Line 13/5/1 - genotype avoided by birds (Stupina)

Figure 24. Polet-11273B - obtained through hybridization backcross x androsterile line. Attack of *Sclerotinia sclerotiorum* on stem, Fundulea (2014).

Figure 25. LC 1093X - obtained through hybridization, with total resistance to *Sclerotinia sclerotiorum*, Fundulea (2014)

Author details

Florentina Sauca* and Catalin Lazar

*Address all correspondence to: tina@ricic.ro

National Agricultural Research and Development Institute (NARDI) – Fundulea, Calaraşi, Romania

References

[1] Vrânceanu, A.V., 2000. Floarea-soarelui hibrida. Editura Ceres, Bucuresti. ISBN 973-40-A0453-0.

[2] Christov, M., 2013. Contribution of interspecific and intergeneric hybridization to sunflower breeding. HELIA 36 (58). ISSN 2197-0483 online. Pg 1-19.

[3] Iuoraş, Monica, Voinescu, A., 1984. The use of the species *Helianthus argophyllus* Torrey and Gray for breeding xerophytic form of sunflower. Probl. Genet. Teor. şi Aplic., 16: 123-130.

[4] Saucă, Florentina, Petcu, E., Stanciu, D., Stanciu, M., 2010. Preliminary identification of Romanian sunflower hybrids suitable for organic agricultural system. RAR. No. 27, 2010; ISSN 2067-5720. Pg 43-46.

[5] www.mapam.ro.

[6] Toncea, I., 2000. Ghid practic de agricultură ecologică; Editura Academicpres. ISBN 973-8266-16-5.

[7] SMART financial - www.SMARTfinancial.ro. Strategia de dezvoltare a agriculturii, industriei alimentare şi silviculturii pe termen lung şi mediu, 2001-2005 şi 2005 - 2010 ; MAAP, 2002).

[8] Report of Organic Agricultural Research Institute, FIBL from Swiss, 2006. www.fibl.org.

[9] The Land Stewardship Centre of Canada and LandWise Inc. www.landstewardship.org/lsn-publications.asp.

7

Organic Livestock Farming — Challenges, Perspectives, and Strategies to Increase Its Contribution to the Agrifood System's Sustainability

Alfredo J. Escribano

Additional information is available at the end of the chapter

Abstract

The livestock sector is of great importance for the sustainability of rural economies and many ecosystems; however, it also has a high environmental impact. Due to the growing demand for animal products, there is a need to design new livestock production systems that allow the combination of food security and sustainability. Within this context, organic livestock may be a useful strategy to achieve such a pivotal goal. However, there is a lack of studies that integrate the existing knowledge, specifically in organic livestock, and integrating the main aspects implied in its practice (its externalities and challenges). The present work aims to fill this knowledge gap, providing strategies and insights that will help stakeholders and policy makers to improve the sustainability of both the organic sector itself and that of the whole food system.

Keywords: Organic, cattle, livestock, sustainability, food system

1. Introduction

There has been considerable growth in the number of organic livestock farms [1] in response to the necessity to fulfill the growing demand for animal products predicted for 2050 [2]. Furthermore, it is required to combine it with the farms' profitability, environmental protection, food safety, and ethical concerns. Due to this, organic livestock farms are nowadays not a despicable part of the census. However, there is no consensus about the consequences of organic livestock farming systems to the sustainability of the overall food system. This lack of convergence has its roots in the effect played by the different characteristics and contexts of the farms. Moreover, some barriers are challenging the development of the sector and shaping its future perspectives. Within this context, and in view of the lack of studies addressing the sustainability of the organic livestock sector as a whole by integrating different points of view, it is very timely to conduct a thorough study of this type. Due to this, the present review was carried out, aimed at improving the knowledge about the organic livestock sector such that it will be possible to adopt a holistic view that increases our understanding of its challenges and future perspectives, with a special emphasis on the sustainability of both farms and the whole food system. This integrative knowledge and approach will help stakeholders and policy makers to make decisions, either at the farm level (implement organic farms) or making policies. Thus, they both will be able to design strategies that increase the sustainability, competitiveness, and success of organic livestock farms, looking at the sustainability of the food systems as a final and priority goal.

1.1. Socio-economic and environmental role of livestock production

Animal production systems are of great importance for the sustainability of rural economies and many ecosystems. The economic importance of livestock activity is reflected by the weight of the agricultural sector in the regional gross domestic product. For example, in rural areas located in southern Europe, cattle, swine, and small ruminants sectors billed 396.46 M € in 2010, representing 36.10% of the agricultural sector production in some regions [3].

From the social point of view, it is noteworthy that in semi-arid regions, such those in the Mediterranean basin, the extensive livestock production systems are often the main activity, and even the only source of livelihood. This dependence of the sector highlights the need to protect and enhance it, as it contributes to the creation of jobs, to the rural economy, and to the fixation of the rural population, which are vital for sustainable development in rural areas worldwide [4-5].

From a cultural perspective, the particularities of the different livestock systems are crucial for the conservation of the heritage, including breeds, landscapes, and habitats of high aesthetic and environmental value [6-7], which redounds in the economic development of the rural areas.

Regarding the environment, livestock activity involves lots of environmental benefits [8], especially when it is carried out under environmentally-friendly production systems, such as the extensive, pasture-based, low-input, and/or organic systems.

However, the livestock sector also has an important environmental impact. This sector employs 30% of the overall area not covered by ice and uses around 80% of global agricultural land. It also generates most of the greenhouse gas (GHGs) emissions in the agricultural sector, accounting for 14.5% of human-induced GHGs, exceeding that from transportation [9].

Moreover, it is a major consumer and polluting water resources, contributing around 30% of N and P content of watercourses [10-13]. These data are even more striking in the case of the bovine meat sector as beef is often the food of animal origin with greater ecological impact [14-18]. Moreover, various socio-economic factors have led to either the abandonment or the intensification of the farms, which threatens the conservation of valuable agro-ecosystems.

Such environmental impact, along with the increasing demand toward animal products, makes it difficult to combine profitability, competitiveness, and environmental sustainability. Consequently, it is necessary to design and implement sustainable livestock systems globally (environmentally, socially, and economically), in which organic ones have an important role to play.

2. Objectives

The objectives of the present work is (i) to fill the existing knowledge gap with regard to the sustainability, challenges, and perspectives of the organic livestock sector, as well as regarding its contribution to the agrifood system's sustainability. Moreover, this study is also aimed at (ii) providing strategies and insights that will help stakeholders and policy makers to improve the sustainability of both the organic livestock sector and that of the whole food system.

3. Externalities of organic livestock farming systems

To reduce the abovementioned environmental impacts, different production systems have been developed. Among them, organic livestock farms have been studied by several authors in order to assess their potential and impact on environmental [19-22] and socio-economic aspects (sustainable rural development) [23-24].

However, some results are contradictory and some papers are not conclusive, which make it difficult to generalize the advantages of the organic livestock sector at either farm level or globally. This lack of convergence in the results is due to the fact that the externalities of organic livestock farms are highly dependent on their structure, the breeds reared, as well as their management, context, and marketing strategies [23-25, 129]. In other words, it seems that there is no one-size-fits-all solution. Moreover, papers normally address specific aspects of the farms (i.e., economic, health, welfare, etc.), which does not allow an integrative picture of the situation.

In order to deal with this scenario, many points must be addressed, as [26] argued, "the concept of organic animal farming can only fulfill the criteria for sustainability if all requirements on animal health, welfare, and ecological soundness are strongly considered and controlled". Due to this, an analysis of the aspects mentioned by these authors, along with those related to the economic and social aspects, have been included in the present work.

3.1. Social dimension: Sustainable rural economy and development

An important part of the world forms part of the so-called "rural areas". In the case of the European Union, rural areas cover 90% of its territory, where over 23% of the European population lives in them, and another 35% lives in intermediate zones. In these areas, farming is one of the main drivers of sustainable rural development [27].

However, these areas are going through processes of depopulation that reduces the sustainability of such areas, from the social, economic, and environmental points of view. Due to this, there is a necessity to develop strategies that allow overcoming this issue. Within these strategies, organic farming has become popular, even in the legislative environment. In fact, [28] defines organic production as "a system of farm management and food production that plays a dual societal role: on the one hand it provides food products to meet specific consumer demands; on the other hand it delivers public goods that contribute to the protection of the environment and animal welfare, as well as to the development of rural areas".

As a consequence, several researchers have evaluated the contribution of organic livestock to sustainable rural development [29-30], of which most of them have been reviewed and discussed by [24]. Some of them have considered that organic production is an important pillar of sustainable rural development, since this production model generates more positive externalities than the conventional one in terms of conservation of agro-ecosystems, creation of jobs, farms' profitability, workers' income, and local economy.

In this regard, it is fair to mention that most of the benefits provided by the organic production model in relation to rural development seem to be due to both their participation in short marketing channels [31-32] and obtaining a higher price ("price premium") for their organic products [33-34]. According to the authors cited, this premium price is necessary for organic farms profitability, especially during the years of conversion, because the farms' incomes are often reduced and costs increased [31-32]. However, there is controversy on the relationship between the condition of being organic and short marketing channels, but in general terms, such relationship is weak [24, 35].

However, few studies addressed the potential role of organic livestock production systems (studies usually mix agriculture and livestock) towards sustainable development, despite having been proposed to be models of it [30]. Furthermore, such studies show contradictory results and did not adopt a holistic approach (social, economic, and environmentally public policies), which is really needed. Due to this, [24] summarized the studies published with regard to organic livestock sector and discussed them with the results of the case study they carried out addressing organic beef cattle farms located in southwestern Europe.

Studies dealing with this topic paid special attention to the number of jobs created, salaries paid, and the profitability of the farms. Thus, authors such as [36] found no increased presence of labor in organic livestock farms when compared with conventional holdings. On the contrary, [24] found that organic beef cattle farms (mainly those that also fattened their calves) used more labor. However, these last authors stated that this was mainly due to both the higher degree of business diversification of these farms and the fact that for many of the farmers, the farm under study was not the only source of income. Both aspects increased the necessity to hire external workforce. Moreover, [24] found that the salaries paid by the organic farms were lower than those of the conventional ones, which is contrary to the findings of the review carried out by [23]. However, these two last studies did not focus only livestock farms, such that results cannot be compared precisely.

[37] concluded that organic dairy farms may contribute more to the local economy and economic development of rural communities located in the northeast and upper midwest of the U.S. than average and similar-size conventional dairy farms. As they stated, in Vermont, organic dairy farm sales revenue would result in greater state-wide impacts of 3% in output, 39% in labor income, 33% in gross state product, and 46% in employment relative to the impacts from an equivalent level of sales revenue to conventional dairy farms. In Minnesota, these economic impacts are 4, 9, 11, and 12% greater.

Later, [24] found that organic beef cattle farms that fattened their calves performed better from the economic point of view. These authors compared these full-cycle organic farms with (i) conventional farms that scarcely fattened their calves and (ii) organic farms with no fattening period. This comparison allowed them to conclude that the differences were mainly due to the consequences that some differential factors had on overall economic performance, more than the condition of being organic. These factors were the following: most of the farms that fattened their calves were full-cycle (they were part of an association that had the organic crops, the mill, the livestock farms, the trucks, and even established contracts with supermarkets). Moreover, they all received the subsidies for organic farming (in the other organic group of farms they did not), and they sold their fattened calves at a price 25% higher than that of the conventional ones. However, as the production cycle (and the age at which calves were slaughtered) was longer, these farms showed lower economic performances when it was calculated per year.

In summary, the authors have come to the conclusion that many of the benefits provided by organic production in relation to rural development are not due to the mere fact produced under the ecological model, but to sell their products through short marketing channels [31-32] and to obtain a higher price ("price premium") for organic products [33-34]. This is especially important during the years of conversion because farm incomes are often reduced, and its costs increased.

Moreover, the pathway followed by the products (marketing channels) has a great impact on the sustainability of the food system. Thus, transportation accounted for 17.43% of the total energy consumed by the Spanish food sector in 2000 [38]. In this sense, it is important to comment that short marketing channels (and "local" products) are commonly thought to have lower environmental impacts. However, the concentration of supply can lead to lower emissions of GHGs of short marketing channels, in which small amounts of products are transported by vehicle or fuel. In fact, [38] found that most of that 17.43% of energy consumed by transport comes from road transport due to their lower energy efficiency per load transported.

3.2. Environmental dimension

Pasture-based and low-input livestock systems (e.g., the organic systems) are key to the ecosystems in which they are integrated as they provide with numerous benefits, such as increased carbon sequestration, improved quality of the pastures, and reduction of scrub invasion and risk of fire [5, 8, 39].

According to [28], livestock production is fundamental to the organization of agricultural production on organic holdings in so far as it provides the necessary organic matter and nutrients for cultivated land and accordingly contributes towards soil improvement and the development of sustainable agriculture. [40] completed this view arguing that organic livestock provides organic nutrients that are recycled at the farm level, allowing the production of on-farm inputs, which increases their sustainability. Similarly, [41] claimed that when cattle are introduced in environmental systems, increased efficiency and sustainability occurs. However, organic livestock farms do not always present a significant cultivated area, so that their differences with conventional farms with regard to this parameter may be few [24]. Moreover, mixed crop-livestock farms could miss out on potential economies of scale. To overcome these interactions, organic mixed crop-livestock farms could be a solution, since [42] observed that these farms exploited the diversity of herd feed resources more efficiently than the rest of the groups, which varied in both their degrees of mixing these two components and their organic/conventional status.

In relation to water resources, some authors have found that its use is more efficient in organic farms, and that water retention is increased, leading to higher resistance to drought [43]. Moreover, in these farms, land degradation is prevented and soil fertility increased [44]. These aspects are of particular interest in semi-arid areas, where water shortages often occur, and both soils and pastures are poor. Additionally, it has been shown that agrobiodiversity is greater in organic agro-ecosystems [20, 21, 45], which greatly increases the number of inter-actions between system components and their complexity. Therefore, their resilience is increased, which is key for their adaptation and resistance against pests, diseases, and climate change. In parallel, their higher degree of business diversification make them less vulnerable in the face of market changes [25, 44, 129].

When looking at comparisons between organic livestock farming systems and conventional ones, several authors have shown that organic systems have a greater potential to preserve the environment, mainly with regard to biodiversity [19-21]. These positive externalities are the consequence of many factors, such as the reduced use of inputs, better nutrient recycling, less use and exploitation of non-renewable/external resources, and finally, ecotoxicity.

These aspects are of great importance, since the increasing degradation of the agricultural soils and the reduction in the supplies of fresh water are two of the most serious problems that humankind is facing. These problems pose an impediment to achieving food security, especially if one takes into account the growing population and demand for animal products. It is even more relevant in developing countries and in semi-arid areas characterized by pasture-based (low-input/pasture-based/extensive) production systems. According to several authors [46-47], organic livestock systems have the potential to contribute to the sustainability of these areas.

Due to the advantages provided by organic livestock production, it would be logical to think that this production model allows facing the two main challenges of the food system: sustainability and food security. In this sense, [48] stated that a shift to organic production will be increasingly necessary for the renewal of resources (mainly water and soil) and to secure sustainable food security. However, there is much debate in this sense [49], due to the lower production that organic production often shows, the increased need for agricultural land for organic production, and the scarcity of organic fertilizers of good quality.

Regarding the environmental impact in terms of GHGs and energy use, extensive and low-input farms (including the organic ones) tend to be more sustainable [50-52]. Among other reasons, this is due to lower consumption of fossil fuels and energy. However, some studies conclude that emissions in organic systems may be higher than those of the conventional ones [16], because they have lower production per unit of input. In this sense, [22] showed that the product carbon footprint in dairy cow organic farms was significantly higher than that of the conventional farms [1.61 ± 0.29 vs. 1.45 ± 0.28 kg of CO_2 equivalents (CO_2 eq) per kg of milk].

This divergent results are showing that the differences among studies are mainly due to the productive system under study, its context, the experimental design work, and the units and limits of the study (farm level, hectare, unit of product, food system, etc.), more than their conditions of being organic.

One of the aspects that plays a great effect on greenhouse emissions of the farms is the quality of the feed. In this sense, [53] measured the GHGs from enteric fermentation and manure on organic and conventional dairy farms in Germany in order to assess the effect of different feeding practices. In general terms, lower emissions from enteric fermentation were found when feed quality and feed intake was increased (which normally means feedstuff, instead of pastures). In general terms, results depended strongly on the calculation methodology, especially those related to enteric fermentation. Moreover, differences between the methods were particularly prominent when high amounts of fiber-rich feedstuff were used. As feed quality management on farms influences milk yield and enteric CH_4 emissions, these aspects should be part of advisory concepts that aim at reducing GHG emissions in milk production.

In line with these results, [22] stated that feed demand per kilogram of milk, high grassland yield, and low forage area requirements per cow are the main factors that decrease PCF (product carbon footprints). They observed that the interaction between GHG mitigation and the farm's profitability is key for improving efficiency and sustainability. Thus, for organic farms, a reduction of feed demand of 100 g/kg of milk resulted in a PCF reduction of 105 g of CO_2 eq/kg of milk and an increase in incomes of approximately 2.1 euro cents (c)/kg of milk. For conventional farms, a decrease of feed demand of 100 g/kg of milk corresponded to a reduction in PCF of 117 g of CO_2 eq/kg of milk and an increase in management incomes MI of approximately 3.1 c/kg of milk. Accordingly, farmers could achieve higher profits while reducing GHG emissions.

Regarding the environmental externalities of the different livestock species and sectors, dairy cows are those that have received more attention. [54] studied the productive, environmental and economic performances of organic and conventional suckler cattle farming systems. They found that the reduction in the use of inputs resulted in a 23% to 45% drop in NRE (non-renewable energy) consumption/ha, 5-20% of which is a drop in non-renewable energy per ton of live weight produced. The authors stated that, however, the shift to organic farming does not significantly affect gross GHG emissions per ton of live weight produced, but suggested that net GHG emissions could be lower for organic farming systems due to the carbon sequestration in grasslands. Contrary to the results that are normally found when GHGs are measured per kg of product, the lower productivity per hectare (fewer animals reared per hectares) allowed a reduction from 26% to 34% in net GHG emissions per hectare of farm area in the study of [54].

[55] reviewed studies that compared different beef production systems using life cycle analysis (LCA). They classified such systems by three main characteristics: origin of calves (bred by a dairy cow or a suckler cow), type of production (organic or non-organic), and type of diet fed to fattening calves (roughage-based -<50% concentrates, or concentrate-based -≥50% concentrates). They observed that organic farms had lowers GWP (global warming potential) and use of energy (on average 7% and 30%, respectively) than that of the non-organic systems. However, they showed higher eutrophization potential, acidification potential, and land use per unit of beef produced. Lower GWP (on average 28% lower), energy use (13% lower), and land use (41% lower) were found per unit of beef for concentrate-based systems when compared with roughage-based systems. Although these results are not giving the whole picture (because aspects such as biodiversity, carbon sequestration, and others were not included in all the studies), the authors came to interesting conclusions that we cite literally:

- Environmental impacts were lower for dairy-based than for suckler-based beef

- GWP was similar for organic and non-organic beef

- GWP, energy use, and land use were lower for concentrate- than roughage-based beef

- Dairy-based beef showed the largest potential to mitigate environmental impacts of beef

- Marginal grasslands unsuitable for dairy farming may be used for production of suckler-based beef to contribute to the availability and access to animal-source food

The study of [56] studied the potential environmental impacts of four different types of organic dairy farms, paying special attention on the farm's structure (the percentage of grassland on total farm area, and feeding intensity). The results showed that farms with high feeding intensity tend to show ecological advantages with regard to their climate impact and their demand for land. On the contrary, low-input farms showed to be better with regard to animal welfare, milk quality, and ammonia losses. But more interestingly, when they assessed the overall environmental index of the farms, low-input and mixed ones showed the best results. Finally, the authors pointed out the necessity of using a wider range of environmental parameters, since results may differ greatly between studies, farms, and systems.

[57] measured the carbon footprint of the organic dairy sector, based on farm data from six European countries. The results showed that the main contributor to the farm's carbon footprint was enteric fermentation, which has much to do with the feed management, as exposed earlier.

To sum up, high-quality feedstuffs reduce enteric methane emissions, and this is important because these emissions account for a high proportion of total GHGs (45% of them in the study of [57]). However, one must keep in mind that the environmental impact of the farms belongs to just one pillar of global sustainability. Hence, with regard to feed, other factors must be taken into account, such as the competence with human food.

Regarding the methodological aspects of the assessment of farm sustainability, it must be remembered that the different parameters, frameworks, and approaches available, as well as the limit of the study and the context of the farms, make it difficult to integrate results and make conclusions. In this sense, [57] stated that the method for calculating the carbon footprint could be improved, since this calculation does not take account of carbon sequestration. This aspect is very important for extensive livestock systems (either organic or not), especially for ruminant ones, since cattle grazing captures 20% of the CO_2 released into the atmosphere by deforestation and agriculture worldwide [58]. If carbon sequestration were included in the evaluations (as done by [25, 129]), extensive farms and sensitive ecosystems would show better results in the evaluations of their environmental impact, which could lead to higher public support, competitiveness, and sustainability.

In relation to the organic beef cattle sector, [25, 129] carried out a comparative assessment of the sustainability of organic and conventional beef cattle farms located in agroforestry systems and rangelands of southwestern Spain. It is worthy to mention that conventional farms where extensive, pasture-based, and low-input; and that all farms had cows, either with presence of a fattening period of the calves or just selling them at the weaning age. These two last productive orientations where selected as they are representative of the sector and the area under study. The results showed that organic farms had a higher overall sustainability, especially with regard to the environmental dimension. In this sense, the authors reported that the agro-ecosystem management (agricultural practices) and farm structures were slightly more environmentally friendly. For example, organic farms tend to implement more measures to reduce erosion and to improve soil fertility, also developing better dung management that avoided nitrogen fluxes and allowed farmers to elaborate compost. Only clear differences where found regarding the use of pesticides, herbicides, and/or mineral fertilizers. This is consistent with the findings of [59] in smaller organic beef cattle farms located in a more humid area (northwestern Spain).

Hence, the presence of an approach and configuration of the farms oriented to organic principles (namely, the environmental systems) found in the study of [25, 129] was really scarce, since the improvement and/or maintenance of the ecosystem did not constitute an important driver nor a motivation of the farmers to run their organic adventure. A higher degree of farmer's engagement and awareness toward the sustainability of the agrifood sector is needed. Specifically, the implementation of such sustainable management practices of the agro-ecosystem, such as diversification (the integration of crops, livestock, and trees), are advisable for sustainable land use management [60, 61] and reduce their carbon footprint [57]. Also, these measures deserve to be taken into account by policy makers due to their positive agro-environmental and socio-economic externalities [24].

With regard to swine, Dourmad et al. (2014) evaluated the environmental impacts (per kg of pig live weight and per ha of land used) of 15 European pig farming systems, comparing them with their conventional counterparts, among other types of farming systems, from which "traditional" was an interesting classification worthy of being mentioned since they account for an important part of the livestock sector and rural economy of many areas. This system was defined as "using very fat, slow-growing traditional breeds and generally outdoor raising of fattening pigs". When looking at the results, one can observe that the main differences were found between the traditional systems and the rest of farms. Environmental impacts were, in general terms, lower for conventional farms, when they were measured by kg of pig produced. Conversely, when expressed per ha of land use, mean impacts were 10% to 60% lower for traditional and organic systems, depending on the impact category. These results are in line with those abovementioned, and as previously explained, they are mainly due to the higher land occupation per kg of product and the longer productive times.

Another important point that [62] mentioned was the effect of the autochthonous breed on the environmental impact of the farms. They stated that the use of traditional local breeds, with reduced productivity and feed efficiency, results in higher impacts per kg of live weight. [63] added that the effects of the use of autochthonous breeds have not been adequately demonstrated with regard to some topics (different than the preservation of the genetic heritage and traditional landscape—aesthetical values). Due to this, [24] and [63] highlighted the necessity to deeply study the interactions and effects of the different livestock systems, especially those with beef cattle, since the scientific literature in addressing this sector is scarce. In line with this argument, [64] mentioned that agricultural practices affect biodiversity in a higher degree than the breeds itself.

Due to these results, context, and the scientific literature available that addressed the topic, [25, 129] came to the conclusion that the externalities of organic farms (when compared with the conventional ones), are highly dependent on their production system, their context (socioeconomic, environmental, political, and institutional), and their marketing strategies. These conclusions can also be found in other studies, such as the review of [23] about the organic sector as a whole and its relationship with rural development.

Therefore, the future strategy of research and innovation in organic farming must prioritize productivity gains that address the farms as a whole, while paying major attention to secure the positive ecological performance organic agriculture can provide, since the environmental benefits it provides are absolute goods and cannot be relativized by the fact that yields are currently lower than in conventional agriculture. Moreover, there is a high potential for reducing the yield gap between organic and conventional farms through agricultural research [47].

4. Factors influencing organic livestock farms' success

4.1. Regulation and certification bodies

With regard to the legislative side, it is very important to note that regulations on organic production embrace a wide variety of organic farms; they allow using different animal breeds, structures, agro-ecosystem managements, feeding strategies, and marketing strategies. As a consequence, organic the livestock farm's success and perspectives are really different from one place to another. For example, [65] found that the situation in North Germany was in contrast to the region in the south, where the variability of amount and proportion of the different feed types is predominantly independent of the milk yield. Many factors shape these differences, such as the ecosystems on which farms are based and consumers' demands and willingness to pay.

Additionally, the different criteria of the certification bodies (public and private) act in the same way, since they usually decide whether some exceptions to the regulations can be applied at the farm level. Due to this, it is important to unify criteria. Also, the cost of certification is not affordable for many farmers (especially small farmers, which play a great role in sustainability and food security). Fortunately, nowadays, many efforts are being made to both facilitate the market of organic products worldwide (i.e., agreements between the European and American (USDA) standards) and to reduce cost of certification (i.e., by means of Participatory Guarantee Systems).

Moreover, organic regulations and private standards do not cover marketing aspects (key in the social, economic, and environmental sustainability), so that it is difficult to evaluate to contribution of the organic livestock sector to the sustainability of the food system.

4.2. Implementation of organic farms: Its consequences on the farms' economic and productive performance

Some studies have assessed the consequences of converting livestock farms to the organic system. Their feasibility and success depend upon the structure and context of the previous (conventional) farm. To cite an example, ruminants pasture-based farms such as those located in southwestern Europe and in the Mediterranean basin (especially those oriented to meat production) may be easily converted into organic ones since conventional and organic farms are quite similar [66-67]. On the contrary, species that are mainly reared under intensive production systems will have to go through a difficult process of conversion, e.g., poultry, swine, and dairy cows. And in parallel depending on the farmers' motivations for converting, the situation of the farms, and their perspectives vary.

As monogastric production systems are not so linked to land as ruminants ones are, and due to the higher prices of organic feedstuffs, it is far more difficult for farmers to convert to produce under the organic system. In this sense, swine rearing under free range production systems (such as those of the dehesa ecosystem in southwestern Spain) appears to be the system that could be converted to the organic model successfully. However, the weaning period seems to be the bottleneck of this sector, because many veterinary interventions are usually needed.

Moving from species to farms structure, it is interesting to note that mixed livestock production systems are those with a higher resilience (also economically), which would allow an easier transition to the organic system [25, 129]. Accordingly, [68] claimed that co-grazing sows with heifers can diminish the parasite burden of the heifers, and that the pig inclination for rooting can be managed in a way that makes ploughing and other heavy land cultivation more or less superfluous. With regard to poultry, there is an indication that quite big flocks can be managed efficiently in a way where the flock act as weeders in other crops or fight pests in orchards. This integration of feed resources of the farms with the different livestock species is possible due to their different grazing habits [69, 70], and is pivotal for the sustainability of the agro-ecosystems and rural areas [25, 129].

However, the consequences of the conversion process and externalities of organic farms may be very changing, since they depend on many factors [66, 23, 25, 35, 129], such as the socioeconomic and environmental context of exploitation, the climate and topography of the land, the production system under study, the species reared, the regulations on course, the influence of private standards of certification, the availability of organic inputs and prices thereof, the development of the organic industry and marketing channels, and the consumer's behavior (demand and willingness to pay). In order to deal with these uncertainties, researchers have conducted studies that have evaluated the ease of conversion of different conventional farming systems to the former one: for dairy goats [71] and dairy cattle [72]. Therefore, before making conclusions about the adequacy of organic livestock farming, one must establish the limits of the study (local or global scale), its objectives, and motivations. Later, a multidisciplinary assessment of farm sustainability, a SWOT analysis, and an assessment of the feasibility of success along with a study of farms competitiveness must be carried out, as proposed by [67, 73].

In relation to organic beef cattle farms, although there is controversy, studies mainly show that organic farms have worse economic results than their conventional counterpart when they are studied by farm and year since they used to have longer production cycles when the farms are under the Common Agricultural Policy's (CAP) conditions [25, 59, 129] or not [74]. They are also more dependent on both subsidies and premium prices. Finally, higher production costs (mainly derived from feeding and during the conversion period) have also been observed [25, 59, 74-75, 129].

[54] analyzed the productive, environmental and economic impacts of the conversion process of conventional suckler cattle farms. They reported that the ban on chemical fertilizers led to a drop in farm area productivity and meat production (by 18% to 37% for the latter) and farm income (more than 20%). These drops were not compensated by the increase in the meat selling price (+5% to +10%). However, the use of inputs was reduced (by -9% to -52%), which is really important for the sustainability of pasture-based/low-input ruminant farms.

With regard to milk production, [76] found that organic systems had greater milk production. However, it seems that milk production per animal [77] and agricultural area [40, 78-79] is lower in organic farms.

Although at first glance, this lower milk production seems negative, this could have very positive implications and advantages. Firstly, cows could have a longer productive life (longevity), which in turn could make animals produce more liters in their entire life, thus reducing the environmental and economic impact of rearing heifers. Secondly, the increase of the productive capacity of the cows has been followed by health problems such as increased somatic cell counts and mastitis, as well as reduced fertility rates and tolerance to heat stress, which could be reduced if cows reduce their production level. Moreover, such reduction would help to reduce the amount and/or proportion of non-structural carbohydrates given to the animals, which would reduce the risk of acidosis, lameness, and other secondary disorders. In this sense, [76] observed that cattle on conventional farms were fed approximately twice as much grain as cattle on organic farms. All these advantages match part of the goals set in the Strategic Research and Innovation Agenda for Organic Food and Farming set by the European Technology Platform (TP Organics) [80]: improved health, robustness, and longevity.

Moreover, as the price of organic milk seems to be more stable [81], the consumption of mothers' milk by calves may be a profitable strategy in farms where milk is not the main marketable product. Thus, [82] found that the consumption of mothers' milk by calves resulted in high weaning weights at 3 months of age, and Keifer et al. (2014) found that organic dairy cows farms performed economically better than the pasture-based conventional farms analyzed.

Not all is about ruminants. Other sectors, such as rabbits, have also been studied. Thus, [83] showed that the effects on zootechnical parameters are due to the production system and genetics. They found that hybrid rabbits reared under conventional housing had the highest average daily gain, and local grey and organic, the lowest.

4.3. Public subsidies: The Common Agricultural Policy (CAP) in the European Union

Despite the abovementioned low productivity in organic farms, their higher environmental externalities should drive a higher support by the rural development measures of the EU's CAP [24, 84-85], since they play a greater role in the conservation of traditional landscapes and ecosystems by means of a "greener" agro-environmental management, which is finally of great importance for the sustainable development of the surrounding rural areas, where the agricultural sector remains an essential driver of the rural development of this area [27]. In this sense, [84] have claimed the necessity to recognize in a higher degree the role of the extensive livestock systems on environmental and cultural heritage preservation.

4.4. Animal nutrition: Legislation and market

Animal nutrition constitutes an important pillar of organic livestock production. Thus, [86] found that feeding strategies among Wisconsin organic dairy farms were major determinants of herd milk production and income over feed costs. These findings may serve current organic and transition farmers when considering feeding management changes needed to meet organic pasture rule requirements or dealing with dietary supplementation challenges.

In relation to organic feedstuffs, the most important obstacles are the difficulty to find them and their prices. This situation is aggravated by the farms' high external dependence of feedstuff due to decoupling between crops and livestock. These facts reduce the organic livestock farms' adaptability, and their access to feed additives and materials of high quality. As a result, the organic livestock sector face a big challenge that, along with other factors, has lead to a situation characterized by organic livestock farms without organic products, which reduces their profitability and future perspectives of success. This has been observed either in beef cattle [25, 129], dairy cows farms [87], or other species [88].

One possible solution for overcoming this barrier would be the use of local agricultural by-products for animal nutrition since their price is usually low, and according to [89], they allow to add to their economic value, while providing an environmentally sound method for disposal of the by-product materials. Also, it would lead to either an increase in the incomes for the organic business that sell such by-products or a reduction in the expenditure related to their disposal.

European regulations limited the use of many feed additives, such as mineral preparations, with the aim that organic livestock farms rely on soil minerals. However, their levels can be low in some areas, which can lead to some mineral deficiencies, as observed by [90] in organic calves. This limitation is especially important in the case of dairy cattle, since nutritional requirements of cows are really high. Due to this, researchers are looking for new feedstuffs that are both allowed and useful for the organic livestock sector, such as minerals sources (seaweed in [91]), different pastures (birdsfoot trefoil by [92]), and fat supplements [93].

As the ration for organic herds has been required to be 100% organic by the European regulations, [94] investigated the possible effects of 100% organic feed on the energy balance in Swedish organic dairy herds as indicated by blood parameters, and concluded that the legislative restrictions "did not appear to have had any detrimental effects on the metabolic profiles of organic cows in early lactation and there was no evidence that organic cows were metabolically more challenged or had a severe negative energy balance".

However, the feed resources of the own farm are usually scarce and/or of poor quality in many areas. Thus, [46] pointed out that the availability of the forages in semi-arid areas, such as the Mediterranean basin, is seasonal, and that its quality is not always optimal. Due to this, the supplementation of the animals is frequently needed. Nevertheless, their availability is low, because for the feed industry it is really costly to turn organic or to create an organic line of products, as they must separate the conventional and the organic lines of productions, and the profitability of this investment is very questionable. Moreover, the bureaucracy would increase the workload of the companies, thus reducing their agility and profitability. In this sense, more concrete instructions for the inclusion of feed additives should be introduced in the regulations.

A correct nutritional management is the basis for an optimal health status and, as a consequence, adequate levels of productivity. Furthermore, this productivity has been identified as key to reduce the GHG emissions from livestock. Due to this, policy makers should seriously address this topic since many conventional companies of the feed sector have a really good portfolio of feed additives that are not susceptible for having not-allowed products (such as GMO or residues of antibiotics), and could improve rumen fermentation (thus reducing the enteric methane emissions), reduce the use of antibiotics (reducing the environmental pollution and public health issues related to them), which would increase the efficiency of the livestock sector, and finally, the competitiveness and sustainability of it. Good examples of additives would be limiting amino acids (such as methionine in dairy cows), chelated (also called "organic") minerals, salts of organic acids, yeasts, essential oils, and fat supplements, among a large list of them. Specifically, organic minerals allow a correct nutritional management, reduce the exploitation of resources, and reduce environmental pollution.

4.5. Animal health, welfare, and technical management

As a consequence of the growth in the number of organic farms worldwide, many veterinarians are encountering this method of production. However, they normally suffer from lack of knowledge with regard to the management of animal health suitable to this type of production, such that it "sustains and enhance the health of soil, plant, animal, human and planet as one and indivisible" (according to IFOAM). The focus is to achieve and maintain high herd health and welfare status with low usage of veterinary medicines [95]. The EC regulations for organic farming [28] state that organic livestock should be treated preferably with phytotherapeutic products. However, almost no phytotherapeutic product is registered for livestock, and information regarding veterinary phytotherapy is really scarce [96].

As health and welfare of organic livestock are highly interrelated, veterinarians not only must avoid livestock illness, but also maintain the animals´ physical, mental, social, and ecological well-being [97]. However, the combination of "natural behavior/living" with optimal health and welfare status is not easy, as [98] and [99] interestingly stated, extensive production systems (e.g., free range production) expose livestock to increased disease challenge, and "a healthy system does not automatically mean good welfare for the individual". However, outdoor housing also has benefits [100]; outdoor housing with functional wallows and access to grass and roots or outdoor runs and roughage can enhance pig welfare and reduce pen-mate-directed oral activity and aggression, which is a really important issue in piglet production.

[99] came to the conclusion that animal health is as good or better than in conventional farming, with the exception of parasitic diseases, and that organic farming systems have a "welfare potential", but organic farmers must deal with the dilemmas and take animal welfare issues seriously. [101] explores how the special organic conceptions of animal welfare are related to the overall principles of organic agriculture. They identified potential routes for future development of organic livestock systems in different contexts (northwestern Europe and tropical low-income countries). Moreover, as outdoor-reared animals make more use of the farm's feed resources, negative consequences can also be found with regard to food safety. Thus, it has been demonstrated that a significant number of organic eggs had dioxin contents that exceeded the EU standard [102].

When one analyzes the health and welfare status of different livestock species, one rapidly realizes that the control of intestinal parasites and to achieve adequate nutritional management are the main bottlenecks and challenges.

Regarding ruminants, [103] also identified these two issues as challenging after studying organic goats. Later, [77] observed lower calf mortality, less incidence of mastitis, fewer rates of spontaneous abortions, and reduced ectoparasite loads in organic farms. However, internal parasite control was again detected as a weak point (greater prevalence was observed in organic farms). Fortunately, animals in the organic system exhibited lower parasitic resistance to anthelmintics, which gives hope to improve herd health status by means of future strategies. [104] reviewed the prevalence of zoonotic or potentially zoonotic bacteria, antimicrobial resistance, and somatic cell counts in organic dairy production; and they found contradictory results in relation with in bacterial outcomes and Somatic Cell Count (SCC) between conventional and organic farms.

Later, [105] discussed the effects of weaning calves at an older age on welfare and milk production. They claimed that foster cow systems with additional milking might be a promising alternative since calves can satisfy their sucking motivation and have social contact to mothers/adult cows; and additionally, weaning stress might be reduced and milking the cows when suckling calves could lead to an increased total milk production. However, this system has economical consequences that must be assessed carefully. Due to this, the authors concluded that further research is needed to reconcile consumers' demands and the possibilities of farmers using such systems.

With regard to animal welfare, [106] assessed the welfare state of dairy cows in European farm systems (extensive and/or low-input farms compared with organic ones) using the Welfare Quality® assessment protocol. Farms had mainly an acceptable and enhanced overall welfare state, although specific problems were found (injuries and discomfort of the lying areas, mutilations, poor human-animal relationship, or insufficient water provision). [107] indicated that most of the organic and conventional farms would have been unlikely to achieve many criteria of audit and assessment programs currently used in the U.S. dairy industry. The parameters recorded were the following: neonatal care, dehorning, pain relief, calf nutrition, weaning, age at weaning, pain relief after and during dehorning, size of the calving area, body condition score, animal hygiene scores, hock lesions, and use of veterinarians. [108] explored how calf welfare is approached in six different organic dairy farms and how far the concept of naturalness is implemented. They observed differing understandings of "naturalness" and welfare, which lead to such diversity of organic farms in aspects that should be shared. In this sense, [82] found that some farmers had difficulties accepting negative implications of suckling systems such as stress after weaning.

The reliance of veterinary drugs is a hot topic that globally is trying to be reduced. In organic farms, where limitations in the use of veterinary drugs are higher, health-related problems can occur, thus undermining the farm's profitability. To reduce these situations, [94], through the CORE Organic ANIPLAN, carried out a study with organic dairy farms of seven European countries, aiming at minimizing medicine use through animal health and welfare planning. Overall, after the implementation of the plan, there was a reduction in the total treatment incidence, and an improvement of the udder health situation across all farms. Hence, these authors concluded that the plan applied "can be regarded as a feasible approach to minimizing medicine use without the impairment of production and herd health under several organic dairy farming conditions in Europe".

Regarding beef cattle, [24, 59] found less use of veterinary medicines. These results are in line with those of [76], who found that the use of outside support and vaccinations were found to be less prevalent on organic dairy farms than on conventional farms. These last authors found little difference in the average reported somatic cell count and standard plate count.

In relation with monogastrics, parasites also constitute a concern. Due to this, the topic was also addressed under the framework of the COREPIG project, a pan-European project on organic pig production focused on the "Prevention of selected diseases and parasites in organic pig herds". One of the results of this project has been the publication of review papers that have provided really valuable information and reflections on the current status and challenges of the swine sector. [109, 110] reported that sows are kept in a variety of different production systems, "with some countries having totally outdoor management at pasture, some keeping animals indoors with concrete outside runs, and others having combinations of these systems". Although reports suggest that relatively few health and welfare problems are seen, the problem of parasites is also a concern within this sector (they are more prevalent in the organic sector). According to the arguments above exposed by [98] and [99], the authors discussed that organic sows had more behavioral freedom, but may be exposed to greater climatic challenges, parasite infestation, and risk of body condition loss. So that, again, the combination of welfare, health, and productivity poses an issue. Even, public health could be compromised, [110] highlighted the high exposure to *T. gondii* in organic pig farms in Italy, indicating a potential risk for meat consumption.

[111] also studied the health and welfare of suckling and weaned piglets in six EU countries. For this purpose, these authors used animal-based parameters from the Welfare Quality® protocol, and showed the main issues prevailing in these farms. [112] studied issues related to weaning in piglets, and they concluded that diseases around weaning are multifactorial so that "in order to solve problems around weaning, the complexity and the individuality of farm systems need to be taken into account".

Furthermore, it has also been reported that some disorders in pigs are less frequent under the organic system, namely, respiratory problems, skin lesions (including abscesses and hernias) and tail wounds. However, joint lesions, white spot livers, and parasitic infections were more common among organic pigs [100]. Due to this, although organic herds consumed three times less antibiotics than conventional ones, the reduction of anthelmintics seems to be more complicated. However, these researchers did not find any difference in mortality rate nor if more pigs in need of treatment in the organic herds.

Fortunately, it seems that some strategies to control the parasites in organic production are coming to scene. Thus, [100] recommended to rotate outdoor areas with as long interval as possible, i.e., by including the pigs in the crop rotation. Furthermore, they stated that an increase in the number of specialized organic farms will help carry out other management strategies needed to maintain the good health of the pigs: implementation of age-segregated production and buying piglets from only one or few units.

Finally, the aquaculture growing sector has also been assessed from the organic side. [113], after studying the open aquaculture systems, reported that both organic and conventional systems present unresolved and significant challenges with regard to the welfare and to environmental integrity, due to many issues such as water quality, escapes, parasites, predator control, and feed-source sustainability. Finally, they concluded that under the current situation, open net-pen aquaculture production cannot be compatible with the principles inherent to organic farming.

4.6. Marketing of organic products and consumer's behavior

Organic livestock farms (when pasture-based and low-input) are perceived as socially more acceptable than intensive ones because they provide many environmental services, such as reducing the risk of fire, improving soil fertility and pastures quality, as well as biodiversity and carbon sequestration. Moreover, they have lower environmental impact linked to land use change (deforestation) and to the use of energy (extraction, manufacturing feedstuff, transportation, etc.) [19-22]. Furthermore, they do not compete with humans for food, which could be another argument to buy organic as the concern about food security has become mainstream. Note that around 70% of the grains used by developed countries are fed to animals and that livestock consume an estimated one-third or more of the world's cereal grain, with 40% of such feed going to ruminants, mainly cattle [114].

However, out of the farm gate, the lack of development of the marketing channels and industry, low consumers awareness of organic products, and their low willingness to pay a premium price for them hinder the demand for organic animal products. As a consequence, most of the farmers are not able to sell their products to the organic market and at a price that allow them to cover their production costs; one can easily find many organic farms without organic products [25, 88, 129]. In the case of livestock, this situation is due to: (i) the difficulty to find organic feedstuff and its cost and (ii) low consumer demand linked to low level of knowledge, awareness, and willingness to pay premium prices. Specifically, in the beef sector, the demand for organic weaned calves (not fattened) was almost non-existent, which make it very difficult to carry out the market of organic beef [25, 129].

In the few cases in which producers can manage to sell their products as organic, such scarcity of developed channels causes the price differential between organic and conventional products to be still high, feeding a loop characterized by reduced per capita consumption and low presence of organic products in the supermarkets [115-116]. As a consequence, demand and willingness to pay consumers for organic products is reduced [117], especially in relation to beef and in countries such as Spain [118-119], despite being one of the first producers in Europe. In order to reduce the cited price differential and increase consumption, a wider distribution of these products is key.

In the case of beef, this little demand is partly due to the fact that consumers do not perceive clearly the differences between organic and conventional meat [115]. Therefore, [120] showed that there is a clear need to excel in organic meat products, quality, and environmental contribution. However, it is can be complex to define and evaluate the quality characteristics of a meat product, especially when the benefits of organic meat over conventional are not clear from the sensory, nutritional, and health aspects[115], particularly when they are compared with conventional extensive systems, such as those present in the pasture.

In summary, it is necessary to note that the demand for organic meat could stagnate due to the following reasons: price differential with conventional meat, inelasticity of demand for this product, and limited knowledge and awareness about the product by consumers. Fortunately, there are strategies that could solve this weak domestic demand, such as exporting. However, meat export is not a strategy easy to carry out due to the cost of transportation and storage, the bureaucracy, and the needed know-how.

Moreover, the approach should not be to just find the markets for organic products, but other additional strategies must be studied. Firstly, it must be taken into account that there is a change in consumer preferences towards local [121-123] and more sustainable [122, 124] products. Moreover, the level of knowledge and awareness about organic products is really low in some countries and regions in Europe [119], leading to the fact that consumers find it hard to differentiate between organic, local, traditional, and sustainable [122, 125-127]. Additionally, one cannot assume that all consumers believe that all organic products are totally complying with the organic principles (many consumers may have not even heard about such principles) and that the organic principles match with the internal triggers and values of the consumers.

To overcome this diversity in the market, organic products should try to be linked to other quality standards. The products with more added value (they would be more than organic) and the growing consumer preferences towards them have both been called 'organic-plus', and have been described by some authors [124]. Within this trend, environmental sustainability, freshness, and local economy are attributes of relevance. In other words, the consequences of the agrifood system (marketing channels, distribution) are becoming important for a growing number of consumers. However, these topics are not covered by the organic regulations, and most of the organic products have been produced and marketed through the mainstream agrifood system; conventional marketing channels characterized by the concentration of production, exporting most of the production, low domestic consumption, and concentration in supply centers and large retail chains. This orientation of organic production into conventional marketing channels and production systems (monocultures and agrochemicals) has been well-documented and is known as "conventionalization" of the organic production and "input substitution" [128].

As a consequence, this type of production (despite being organic) does not always provide consumers with products as fresh, local, and sustainable as they desire, nor positively impact environmental protection and/or rural development in such degree, as was explained above.

In summary, it seems that organic products are not the solution for many consumers that really want to access sustainable products. If organic companies and/or policy makers do not take into account these aspects, the growth of the organic sector, as well as their positive externalities, will be limited.

5. Conclusions

Organic livestock farming (especially its organic principles than regulations) may be a useful strategy to overcome the challenges of the agricultural sector (sustainability, food security, and food safety) while matching with consumers' tendencies (animal welfare, health, environmental protection, etc.). Furthermore, organic livestock farming could be also an interesting strategy for the eternal rural development issue and the farms' decreasing profitability.

However, the combination of complying with organic regulations and objectives and principles of organic farming while increasing overall sustainability is not an easy task. Due to this, it is inappropriate to generalize the benefits of organic livestock farming itself, since the feasibility of implementing organic livestock production systems and their consequences varies greatly, and are site and time-specific. Therefore, it must be remembered that any production system that does not evolve from its initial state (i.e., defined by law) and do not take into account both the time and spatial scales cannot be sustainable worldwide and for a long time. Due to this, a SWOT study along with an assessment of the future effects and difficulties of organic farms under specific contexts is really needed. By doing so, it will be possible to design site-specific and successful options that comply with organic regulations and principles, while being sustainable.

Moreover, some topics must be addressed in order to increase the organic livestock farm's success. Firstly, it has been observed that most of the farmers do not focus on sustainability nor environmental improvement, and that many farms are easily complying with the organic regulations without carrying out environmentally-friendly management practices in their agro-ecosystems. Due to this, improved education and training of farmers and consultants regarding conservation agriculture and GHG mitigation are really needed.

Secondly, there is a need to design feeding strategies that provide adequate nutrition, especially in areas with environmental constraints, such as arid and semi-arid areas. Moreover, regulations should both unify criteria and facilitate the production of feed additives by companies, because the consequences of it could be really important and positive for the organic livestock sector and for the sustainability of the food system.

Thirdly, the knowledge of the veterinarians with regard to animal health management must be improved as fast as the sector is growing. Related to this, more light must be shed on the relationship between animal welfare, "natural living-behavior", and animal health. Furthermore, health care protocols must be developed for each species, including research on alternative and complementary methods of disease prevention.

Fourthly, CAP schemes should be improved in order to reward systems that produce positive externalities in a greater extent despite being low in productivity, since the agricultural sector remains an essential driver of rural areas. These systems contribute to environmental, cultural, and heritage conservation, which finally lead to revitalized rural areas and overall sustainability (from the economic, social, and environmental standpoints).

Finally, and more urgently, special attention must be paid on the marketing strategies of organic products (organic plus products and marketing channels) since this is the main constraint of the sector, and it is the point where there are more possibilities for improvement for both farm profitability and overall sustainability of the food system.

Author details

Alfredo J. Escribano

Address all correspondence to: ajescc@gmail.com

Researcher and consultant. C/ Rafael Alberti, Cáceres, Spain

References

[1] FiBL, IFOAM. The World of Organic Agriculture. Statistics & emerging trends 2015. Frick and Boon: FiBL and IFOAM. 303 p.

[2] FAO. The State of Food Insecurity in the World. Economic crises—impacts and lessons learned. Rome: FAO; 2009. 56 p.

[3] Sánchez J. Las macromagnitudes agrarias. In: La Agricultura y la ganadería extremeñas. Informe 2012. Badajoz: Caja de Badajoz; 2013. p. 37-52.

[4] Boyazoglu J, Hatziminaoglou I, Morand-Fehr P. The role of the goat in society: Past, present and perspectives for the future. Small Ruminant Research. 2005;60:13-23. DOI: 10.1016/j.smallrumres.2005.06.003.

[5] De Rancourt M, Fois N, Lavín MP, Tchakérian E, Vallerand F. Mediterranean sheep and goats production: An uncertain future. Small Ruminant Research. 2006;62:167-179. DOI: 10.1016/j.smallrumres.2005.08.012.

[6] Gellrich M, Baur P, Koch B, Zimmermann NE. Agricultural land abandonment and natural forest re-growth in the Swiss mountain: A spatially explicit economic analysis. Agriculture, Ecosystems and Environment. 2007;118:93-108. DOI: 10.1016/j.agee.2006.05.001.

[7] Cocca G, Sturaro E, Gallo L, Ramanzin M. Is the abandonment of traditional livestock farming systems the main driver of mountain landscape change in Alpine areas? Land Use Policy. 2012;29:878-886. DOI: 10.1016/j.landusepol.2012.01.005.

[8] Henkin Z, Ungar ED, Dvash L, Perevolotsky A, Yehuda Y, Sternberg M, Voet H, Landau SY. Effects of cattle grazing on herbage quality in a herbaceous Mediterranean rangeland. Grass and Forage Science. 2011;66:516-525. DOI: 10.1111/j.1365-2494.2011.00808.

[9] Eisler MC, Lee MRF, Tarlton JF, Martin GB, Beddington J, Dungait JAJ, Greathead H, Liu J, Mathew S, Miller H, Misselbrook T, Murray P, Vinod VK, Van Saun R, Winter M. Agriculture: Steps to sustainable livestock. Nature. 2015;507:32-34. DOI: 10.1038/507032a.

[10] FAO. Livestock's Long Shadow. Environmental issues and options. Rome: FAO; 2006. 26 p.

[11] Garnett T. Where are the best opportunities for reducing greenhouse gas emissions in the food system (including the food chain)? Food Policy. 2011;36:523-532. DOI: 10.1016/j.foodpol.2010.10.010.

[12] Ridoutt BG, Sanguansri P, Freer M, Harper GS. Water footprint of livestock: Comparison of six gepgraphically defined beef production systems. The International Journal of Life Cycle Assessment. 2012;17:165-175. DOI: 10.1007/s11367-011-0346-y.

[13] Bellarby J, Tirado R, Leip A, Weiss F, Lesschen JP, Smith P. Livestock greenhouse gas emissions and mitigation potential in Europe. Global Change Biology. 2013,19:3-18. DOI: 10.1111/j.1365-2486.2012.02786.x.

[14] Baroni L, Cenci L, Tettamanti M, Berati M. Evaluating the environmental impact of various dietary patterns combined with different food production systems. European Journal of Clinical Nutrition. 2007;61:279-286. DOI: 10.1038/sj.ejcn.1602522.

[15] De Vries M, de Boer IJM. Comparing environmental impacts for livestock products: A review of life cycle assessments. Livestock Science. 2010;128:1-11. DOI: 10.1016/j.livsci.2009.11.007.

[16] Lynch DH, MacRae R, Martin RC. The carbon and Global Warming Potential impacts of organic farming: Does it have a significant role in an energy constrained world? Sustainability. 2011;3:322-362. DOI: 10.3390/su3020322.

[17] Dumortier J, Hayes DJ, Carriquiry M, Dong F, Du X, Elobeid A, Fabiosa JF, Martin PA, Mulik K. The effects of potential changes in United States beef production on global grazing systems and green house gas emissions. Environmental Research Letters. 2012;7:1-9.

[18] Mancini L, Lettenmeier M, Rohn H, Liedtke C. Application of the MIPS method for assessing the sustainability of production-consumption systems of food. Journal of Economic Behaviour & Organization. 2012;81:779-793. DOI: 10.1016/j.jebo.2010.12.023.

[19] Gomiero T, Pimentel D, Paolettia MG. Environmental impact of different agricultural management practices: Conventional vs. organic agriculture. Critical Reviews in Plant Sciences. 2011;30:95-124. DOI: 10.1080/07352689.2011.554355.

[20] Halberg N. Assessment of the environmental sustainability of organic farming: Definitions, indicators and the major challenges. Canadian Journal of Plant Science. 2012;92:981-996. DOI: 10.1079/PAVSNNR20127010.

[21] Tuomisto HL, Hodge ID, Riordan P, Macdonald DW. Does organic farming reduce environmental impacts? A meta-analysis of European research. Journal of Environmental Management. 2012;112:309-320. DOI: 10.1016/j.jenvman.2012.08.018.

[22] Kiefer L, Menzel F, Bahrs E. The effect of feed demand on greenhouse gas emissions and farm profitability for organic and conventional dairy farms. Journal of Dairy Science. 2014;97:7564-7574. DOI: 10.3168/jds.2014-8284.

[23] Lobley M, Butler A, Reed M. The contribution of organic farming to rural development: An exploration of the socio-economic linkages of organic and non-organic farms in England. Land Use Policy. 2009;26:723-735. DOI: 10.1016/j.landusepol. 2008.09.007.

[24] Escribano AJ, Gaspar P, Mesias FJ, Escribano M. The contribution of organic livestock to sustainable rural development in sensitive areas. International Journal of Research Studies in Agricultural Sciences (IJRSAS). 2015;1:21-34.

[25] Escribano AJ, Gaspar P, Mesias FJ, Pulido AF, Escribano M. A sustainability assessment of organic and conventional beef cattle farms in agroforestry systems: The case of the dehesa rangelands. ITEA. 2014a;110:343-367. DOI: 10.12706/itea.2014.022.

[26] Von Borell E, Sørensen JT. Organic livestock production in Europe: Aims, rules and trends with special emphasis on animal health and welfare. Livestock Production Science. 2004;90:3-9. DOI: 10.1016/j.livprodsci.2004.07.003.

[27] Manos B, Bournaris T, Chatzinikolaou P, Berbel J, Nikolov D. Effects of CAP policy on farm household behaviour and social sustainability. Land Use Policy. 2013;31:166-181. DOI: 10.1016/j.landusepol.2011.12.012.

[28] Council Regulation (EC) No. 834/2007 of 28 June 2007 on organic production and labelling of organic products and repealing Regulation (ECC) No. 2092/91.

[29] Espinoza-Villavicencio JL, Palacios-Espinosa A, Ávila-Serrano N, Guillén-Trujillo A, De Luna-De La Peña R, Ortega-Pérez R, Murillo-Amador B. Organic livestock, an alternative of cattle development for some regions of Mexico: A review. Interciencia. 2007;32:6.

[30] Pauselli M. Organic livestock production system as a model of sustainable development. Italian Journal of Animal Science. 2009;8:581-587. DOI.org/10.4081%2Fijas. 2009.s2.581.

[31] Seyfang G. Ecological citizenship and sustainable consumption: Examining local organic food networks. Journal of Rural Studies. 2006;22:383-395. DOI: 10.1016/j.jrurstud.2006.01.003.

[32] Wittman H, Beckie M, Hergesheimer C. Linking local food systems and the social economy? Future roles for farmers' markets in Alberta and British Columbia. Rural Sociology. 2012;77:36-61. DOI: 10.1111/j.1549-0831.2011.00068.x.

[33] Tzouramani I, Sintori A, Liontakis A, Karanikolas P, Alexopoulos G. An assesssment of the economic performance of organic dairy sheep farming in Greece. Livestock Science. 2011;141:136-142. DOI: 10.1016/j.livsci.2011.05.010.

[34] Sahm H, Sanders J, Nieberg H, Behrens G, Kuhnert H, Strohm R, Hamm U. Reversion from organic to conventional agriculture: A review. Renewable Agriculture and Food Systems. 2013;28:263-275. DOI: 10.1017/S1742170512000117.

[35] Lobley M, Butler A, Winter M. Local organic food for local people? Organic marketing strategies in England and Wales. Regional Studies. 2013;47:216-228. DOI: 10.1080/00343404.2010.546780.

[36] Butler L. Survey quantifies cost of organic milk production in California. California Agriculture. 2002;56:157-162. DOI: 10.3733/ca.v056n05p157.

[37] O'Hara JK, Parsons RL. The economic value of organic dairy farms in Vermont and Minnesota. Journal of Dairy Science. 2013;96:6117-6126. DOI: 10.3168/jds.2013-6662.

[38] Infante Amate J, González de Molina M. 'Sustainable de-growth' in agriculture and food: An agro-ecological perspective on Spain's agri-food system (year 2000). Journal of Cleaner Production. 2011;38:1-9. DOI: 10.1016/j.jclepro.2011.03.018.

[39] Jose S. Agroforestry for ecosystem services and environmental benefits: An overview. Agroforestry Systems. 2009;34:27-31. DOI: 1 0.1007/s10457-009-9229-7.

[40] Nemecek T, Huguenin-Elie O, Dubois D, Gaillard G, Schaller B, Chervet A. Life cycle assessment of Swiss farming systems, II. Extensive and intensive production. Agricultural Systems. 2011;104:233-245. DOI: 10.1016/j.agsy.2010.07.007.

[41] Gaudino S, Goia I, Borreani G, Tabacco E, Sacco D. Cropping system intensification grading using an agro-environmental indicator set in northern Italy. Ecological Indicators. 2014;40:76-89. DOI: 10.1016/j.ecolind.2014.01.004.

[42] Veysset P, Lherm M, Bébin D, Roulenc M. Mixed crop-livestock farming systems: A sustainable way to produce beef? Commercial farms results, questions and perspectives. Animal. 2014;8:1218-1228. DOI: 10.1017/S1751731114000378.

[43] Thierfelder C, Wall PC. Effects of conservation agriculture techniques on infiltration and soil water content in Zambia and Zimbabwe. Soil and Tillage Research. 2009;105:217-227. DOI: 10.1016/j.still.2009.07.007.

[44] Niggli U, Early J, Ogorzalek K. Organic agriculture and environmental stability of the food supply. In: Proccedings of the International Conference on Organic Agriculture and Food Security. Rome: FAO; 2007. 20 p.

[45] Gomiero T, Pimentel D, Paolettia MG. Environmental impact of different agricultural management practices: Conventional vs. organic agriculture. Critical Reviews in Plant Sciences. 2011;30:95-124. DOI: 10.1080/07352689.2011.554355.

[46] Ronchi B, Nardone A. Contribution of organic farming to increase sustainability of Mediterranean small ruminants livestock systems. Livestock Production Science. 2003;80:17-31. DOI: 10.1016/S0301-6226(02)00316-0.

[47] Niggli U. Sustainability of organic food production: Challenges and innovations. Proceedings of the Nutrition Society. 2014;1-4. DOI:10.1017/S0029665114001438.

[48] Azadi H, Schoonbeek S, Mahmoudi H, Derudder B, De Maeyer P, Witlox F. Organic agriculture and sustainable food production system: Main potentials. Agriculture, Ecosystems and Environment. 2011;144:92-94. DOI: 10.4141/cjss2013-095.

[49] Godfray HCJ, Beddington JR, Crute IR, Haddad L, Lawrence D, Muir JF, Pretty J, Robinson S, Thomas SM, Toulmin C. Food security: The challenge of feeding 9 billion people. Science. 2010;327:812-818. DOI: 10.1126/science.1185383.

[50] Casey JW, Holden NM. Greenhouse gas emissions from conventional, agri-environmental scheme, and organic Irish suckler-beef units. Journal of Environmental Quality. 2006;35:231-239. DOI: 10.2134/jeq2005.0121.

[51] Mondelaers K, Aertsens J, Van Huylenbroeck G. A meta-analysis of the differences in environmental impacts between organic and conventional farming. British Food Journal. 2009;111:1098-1119.

[52] Boggia A, Paolotti L, Castellini C. Environmental impact evaluation of conventional, organic and organic-plus poultry production systems using life cycle assessment (review). World's Poultry Science Journal. 2010;66:95-114. DOI: 10.1017/S0043933910000103.

[53] Warnecke S, Paulsen HM, Schulz F, Rahmann G. Greenhouse gas emissions from enteric fermentation and manure on organic and conventional dairy farms - an analysis based on farm network data. Organic Agriculture. 2014;(4):285-293. DOI: 10.1007/s13165-014-0080-4.

[54] Veysset P, Lherm M, Bébin D. Productive, environmental and economic performances assessments of organic and conventional suckler cattle farming systems. Organic Agriculture. 2011;1:1-16. DOI: 10.1007/s13165-010-0001-0.

[55] De Vries M, van Middelaar CE, de Boer IJM. Comparing environmental impacts of beef production systems: A review of life cycle assessments. DOI: 10.1016/j.livsci.2015.06.020.

[56] Müller-Lindenlauf M, Deittert C, Köpke U. Assessment of environmental effects, animal welfare and milk quality among organic dairy farms. Livestock Science. 2010;128:140-148.

[57] Hietala S, Smith L, Knudsen MT, Kurppa S, Padel S, Hermansen JE. Carbon footprints of organic dairying in six European countries—real farm data analysis. Organic Agriculture. 2015;5:91-100. DOI 10.1007/s13165-014-0084-0.

[58] Ronald FF, Debbie AR. Soil carbon sequestration in grazing lands: Societal benefits and policy implications. Rangeland Ecology & Managament. 2010;63:4-15. DOI: 10.2111/08-225.1.

[59] Blanco-Penedo I, López-Alonso M, Shore RF, Miranda M, Castillo C, Hernández J, Benedito JL. Evaluation of organic, conventional and intensive beef farm systems: Health, management and animal production. Animal. 2012a;6:1503-1511. DOI: 10.1017/S1751731112000298.

[60] Smith J, Pearce BD, Wolfe MS. Reconciling productivity with protection of the environment: Is temperate agroforestry the answer? Renewable Agriculture and Food Systems. 2013;28:80-92. DOI: 10.1017/S1742170511000585.

[61] Cook SL, Ma Z. Carbon sequestration and private rangelands: Insights from Utah landowners and implications for policy development. Land Use Policy. 2014;36:522-532. DOI: 10.1016/j.landusepol.2013.09.021.

[62] Dourmad JY, Ryschawy J, Trousson T, Bonneau M, Gonzàlez J, Houwers HW, Hviid M, Zimmer C, Nguyen TL, Morgensen L. Evaluating environmental impacts of contrasting pig farming systems with life cycle assessment. Animal. 2014;8:2027-37. DOI: 10.1017/S1751731114002134.

[63] Morgan-Davies J, Morgan-Davies C, Pollock ML, Holland JP, Waterhouse A. Characterisation of extensive beef cattle systems: Disparities between opinions, practice and policy. Land Use Policy. 2014;38:707-718. DOI: 10.1016/j.landusepol.2014.01.016.

[64] Scimone M, Rook AJ, Garel JP, Sahin N. Effects of livestock breed and grazing intensity on grazing systems: 3. Effects on diversity of vegetation. Grass and Forage Science. 2007;62:172-184. DOI: 10.1111/j.1365-2494.2007.00579.x.

[65] Haas G, Deittert C, Köpke U. Impact of feeding pattern and feed purchase on area- and cow-related dairy performance of organic farms. Livestock Science. 2007;106:132-144. DOI: 10.1016/j.livsci.2006.07.007.

[66] Nardone A, Zervas G, Ronchi B. Sustainability of small ruminant organic systems of production. Livestock Production Science. 2004;90:27-39. DOI: 10.1016/j.livprodsci.2004.07.004.

[67] Escribano AJ, Gaspar P, Mesias FJ, Escribano M, Pulido AF. In: Proccedings of the Annual Meeting of the European Association for Animal Production (65th EAAP); 25-29 August 2014; Denmark. Copenhagen: EAAP, 2014c. 248 p.

[68] Hermansen JE, Strudsholm K, Horsted K. Integration of organic animal production into land use with special reference to swine and poultry. Livestock Production Science. 2004;90:11-26. DOI: 10.1016/j.livprodsci.2004.07.009.

[69] Anderson DM, Fredrickson EL, Estell RE. Managing livestock using animal behaviour: Mixed-species stocking and flerds. Animal. 2012;6:1339-1349. DOI: 10.1017/S175173111200016X.

[70] Ferraz de Oliveira MI, Lamy E, Bugalho MN, Vaz M, Pinheiro C, Cancela d'Abreu M, Capela e Silva F, Sales-Baptista E. Assessing foraging strategies of hervibores in Med-

iterranean oak woodlands: A review of key issues and selected methodologies. Agroforestry Systems. 2013;87:1421-1437. DOI: 10.1007/s10457-013-9648-3.

[71] Mena Y, Nahed J, Ruiz FA, Sánchez-Muñoz JB, Ruiz-Rojas JL, Castel JM. Evaluating mountain goat dairy systems for conversion to the organic model, using a multicriteria method. Animal. 2012;6:693-703. DOI: 10.1017/S175173111100190X.

[72] Nahed-Toral J, Sánchez-Muñoz B, Mena Y, Ruiz-Rojas J, Aguilar-Jimenez R, Castel JM, de Asis Ruiz F, Orantes-Zebadua M, Manzur-Cruz A, Cruz-López J, Delgadillo-Puga C. Feasibility of converting agrosilvopastoral systems of dairy cattle to the organic production model in south eastern Mexico. Journal of Cleaner Production. 2013;43:136-145. DOI: 10.1016/j.jclepro.2012.12.019.

[73] Escribano AJ, Gaspar P, Mesias FJ, Escribano M, Pulido AF. Competitiveness of extensive beef cattle farms located in the dehesa ecosystem (SW Europe). In: Proccedings of the Annual Meeting of the European Association for Animal Production (65th EAAP); 25-29 August 2014; Denmark. Copenhagen: EAAP; 2014b. 263 p.

[74] Gillespie J, Nehring R. Comparing economic performance of organic and conventional U.S. beef farms using matching samples. Australian Journal of Agricultural and Resource Economics. 2013;57:178-192. DOI: 10.1111/j.1467-8489.2012.00610.x.

[75] Esterhuizen J, Groenewald IB, Strydom PE, Huego A. The performance and meat quality of Bonsmara steers raised in a feedlot, on conventional pastures or on organic pastures. South African Journal of Animal Sciences. 2008;38:303-314. DOI: 10.4236/jep.2011.24046.

[76] Stiglbauer KE, Cicconi-Hogan KM, Richert R, Schukken YH, Ruegg PL, Gamroth M. Assessment of herd management on organic and conventional dairy farms in the United States. Journal of Dairy Science. 2013;96:1290-1300. DOI: 10.3168/jds.2012-5845.

[77] Silva JB, Fagundes GM, Soares JPG, Fonseca AH, Muir JP. A comparative study of production performance and animal health practices in organic and conventional dairy systems. Tropical Animal Health and Production. 2014;46:1287-1295. DOI: 10.1007/s11250-014-0642-1.

[78] Leifeld J. How sustainable is organic farming? Agriculure, Ecosystems and Environment. 2012;150:121-122. DOI: 10.1007/s13280-010-0082-8.

[79] De Ponti T, Rijk B, Van Ittersum MK. The crop yield gap between organic and conventional agriculture. Agricultural Systems. 2012;108:1-9. DOI: 10.1016/j.agsy.2011.12.004.

[80] TP Organics. 2014. Strategic Research and Innovation Agenda for Organic Food and Farming. TP organic. Brussels: TP Organics; 2014. 60 p.

[81] Lebacq T, Baret PV, Stilmant D. Role of input self-sufficiency in the economic and environmental sustainability of specialised dairy farms. Animal. 2015;9:544-552. DOI: 10.1017/S1751731114002845.

[82] Wagenaar JPTM, Langhout J. Practical implications of increasing 'natural living' through suckling systems in organic dairy calf rearing. NJAS - Wageningen Journal of Life Sciences. 2007;54:375-386. DOI: 10.1016/S1573-5214(07)80010-8.

[83] Paci G, Zotte AD, Cecchi F, De Marco M, Schiavone A. The effect of organic vs. conventional rearing system on performance, carcass traits and meat quality of fast and slow growing rabbits. Animal Science Papers and Reports. 2014;32:337-349.

[84] Franco JA, Gaspar P, Mesías FJ. Economic analysis of scenarios for the sustainability of extensive livestock farming in Spain under the CAP. Ecological Economics. 2012;74:120-129. DOI: 10.1016/j.ecolecon.2011.12.004.

[85] Gómez-Limón JA, Picazo-Tadeo AJ, Reig-Martínez E. Eco-efficiency assessment of olive farms in Andalusia. Land Use Policy. 2013;29:395-406. DOI: /10.1016/j.landusepol.2011.08.004.

[86] Hardie CA, Wattiaux M, Dutreuil M, Gildersleeve R, Keuler NS, Cabrera VE. Feeding strategies on certified organic dairy farms in Wisconsin and their effect on milk production and income over feed costs. Journal of Dairy Science. 2014;97:4612-4623. DOI: 10.3168/jds.2013-7763.

[87] Flaten O, Lien G, Ebbesvik M, Koesling M, Valle PS. Do the new organic producers differ from the 'old guard'? Empirical results from Norwegian dairy farming. Renewable agriculture and Food Systems. 2006;21:174-182. DOI: 10.1079/RAF2005140.

[88] Argyropoulos C, Tsiafouli MA, Sgardelis SP, Pantis JD. Organic farming without organic products. Land Use Policy. 2013;32:324-328. DOI: 10.1016/j.landusepol.2012.11.008.

[89] Rinne M, Dragomir C, Kuoppala K, Smith J. Yáñez-Ruiz D. Novel feeds for organic dairy chains. Organic Agriculture. 2014;4:275-284. DOI: 10.1007/s13165-014-0081-3.

[90] Blanco-Penedo I, Shore RF, Miranda M, Benedito JL, López-Alonso M. Factors affecting trace element status in calves in NW Spain. Livestock Science. 2009;123:198-208. DOI: 10.1016/j.livsci.2008.11.011.

[91] Rey-Crespo F, López-Alonso M, Miranda M. The use of seaweed from the Galician coast as a mineral supplement in organic dairy cattle. Animal. 2014;8:580-586. DOI: 10.1017/S1751731113002474.

[92] Hunt SR, MacAdam JW, Reeve JR. Establishment of birdsfoot trefoil (Lotus corniculatus) pastures on organic dairy farms in the Mountain West USA. Organic Agriculture. 2015;5:63-77. DOI: 10.1007/s13165-014-0091-1.

[93] Johansson B, Kumm K-I, Åkerlind M, Nadeau E. Cold-pressed rapeseed cake or full fat rapeseed to organic dairy cows—milk production and profitability. Organic Agriculture. 2015;5:29-38. DOI: 10.1007/s13165-014-0094-y.

[94] Blanco-Penedo I, Fall N, Emanuelson U. Effects of turning to 100% organic feed on metabolic status of Swedish organic dairy cows. Livestock Science. 2012b; 143:242-248. DOI: 10.1017/S1751731112000298.

[95] Ivemeyer S, Smolders G, Brinkmann J, Gratzer E, Hansen B, Henriksen BIF, Huber J, Leeb C, March S, Mejdell C, Nicholas P, Roderick S, Stöger E, Vaarst M, Whistance LK, Winckler C, Walkenhorst M. Impact of animal health and welfare planning on medicine use, herd health and production in European organic dairy farms. Livestock Science. 2012;145:63-72. DOI: 10.1016/j.livsci.2011.12.023.

[96] Mayer M, Vogl CR, Amorena M, Hamburger M, Walkenhorst M. Treatment of organic livestock with medicinal plants: A systematic review of European ethnoveterinary research. Forsch Komplementmed. 2014;21:375-386. DOI: 10.1159/000370216.

[97] Edwards SA, Prunier A, Bonde M, Stockdale EA. Special issue—organic pig production in Europe—animal health, welfare and production challenges. Organic agriculture. 2014a;4:79-81. DOI: 10.1007/s13165-014-0078-y.

[98] Vaarst M, Padel S, Hovi M, Younie D, Sundrum A. Sustaining animal health and food safety in European organic livestock farming. Livestock Production Science. 2005;94:61-69. DOI: 10.1016/j.livprodsci.2004.11.033.

[99] Lund V. Natural living—a precondition for animal welfare in organic farming. Livestock Science. 2006;100:71-83. DOI: 10.1016/j.livsci.2005.08.005.

[100] Lindgren K, Bochicchio D, Hegelund L, Leeb C, Mejer H, Roepstorff A, Sundrum A. Animal health and welfare in production systems for organic fattening pigs. Organic Agriculture. 2014;4:135-147. DOI: 10.1007/s13165-014-0069-z.

[101] Vaarst M, Alrøe HF. Concepts of animal health and welfare in organic livestock systems. Journal of Agricultural and Environmental Ethics. 2012;25:333-347. DOI: 10.1007/s10806-014-9512-0.

[102] De Vries M, Kwakkel RP, Kijlstra A. Dioxins in organic eggs: A review. NJAS. 2006;54:207-222. DOI: 10.1016/S1573-5214(06)80023-0.

[103] Lu CD, Gangyi X, Kawas JR. 2010. Organic goat production, processing and marketing: Opportunities, challenges and outlook. Small Ruminant Research. 89:102-109.

[104] Wilhelm B, Rajić A, Waddell L, Parker S, Harris J, Roberts KC, Kydd R, Greig J, Baynton A. Prevalence of zoonotic or potentially zoonotic bacteria, antimicrobial resistance, and somatic cell counts in organic dairy production: Current knowledge and research gaps. Foodborne Pathogens and Disease. 2009;6:525-539. DOI: 10.1089/fpd.2008.0181.

[105] Kälber T, Barth K. 2014. Practical implications of suckling systems for dairy calves in organic production systems - A review. Landbauforschung Volkenrode. 2014;64(1): 45-58. DOI: 10.3220/LBF_2014_45-58.

[106] Kirchner MK, Ferris C, Abecia L, Yanez-Ruiz DR, Pop S, Voicu I, Dragomir C, Winckler C. Welfare state of dairy cows in three European low-input and organic systems. Organic Agriculture. 2014;4:309-311. DOI: 10.1007/s13165-014-0074-2.

[107] Bergman MA, Richert RM, Cicconi-Hogan KM, Gamroth MJ, Schukken YH, Stiglbauer KE, Ruegg PL. Comparison of selected animal observations and management practices used to assess welfare of calves and adult dairy cows on organic and conventional dairy farms. Journal of Dairy Science. 2014;97:4269-4280. DOI: 10.3168/jds. 2013-7766.

[108] Vetouli T, Lund V, Kaufmann B. Farmers' attitude towards animal welfare aspects and their practice in organic dairy calf rearing: A case study in selected nordic farms. Journal of Agricultural and Environmental Ethics. 2014;25:349-364. DOI: 10.1007/s10806-010-9301-3.

[109] Edwards S, Mejer H, Roepstorff A, Prunier A. Animal health, welfare and production problems in organic pregnant and lactating sows. Organic Agriculture. 2014b; 4:93-105. DOI: 10.1007/s13165-014-0061-7.

[110] Bacci C, Vismarra A, Mangia C, Bonardi S, Bruini I, Genchi M, Kramer L, Brindani F. Detection of Toxoplasma gondii in free-range, organic pigs in Italy using serological and molecular methods. International Journal of Food Microbiology. 2015,6:54-56. DOI: 10.1016/j.ijfoodmicro.2015.03.002.

[111] Dippel S, Leeb C, Bochicchio D, Bonde M, Dietze K, Gunnarsson S, Lindgren K, Sundrum A, Wiberg S, Winckler C, Prunier A. Health and welfare of organic pigs in Europe assessed with animal-based parameters. Organic Agriculture. 2013;4:149-161. DOI: 10.1007/s13165-013-0041-3.

[112] Leeb C, Hegelund L, Edwards S, Mejer H, Roepstorff A, Rousing T, Sundrum A, Bonde M. Animal health, welfare and production problems in organic weaner pigs. Organic Agriculture. 2014;4:123-133. DOI: 10.1007/s13165-013-0054-y.

[113] Cottee SY, Petersan P. Animal welfare and organic aquaculture in open systems. Journal of Agricultural and Environmental Ethics. 2009;22:437-461. DOI: 10. 1007/s10806-009-9169-2.

[114] FAO. Towards 2015/2030. Rome: FAO; 2002. 97 p.

[115] Krystallis A, Arvanitoyannis I, Chryssohoidis G. Is there a real difference between conventional and organic meat? Investigating consumers' attitudes towards both meat types as an indicator of organic meat's market potential. Journal of Food Products Marketing. 2006;12:47-78. DOI: 10.1300/J038v12n02_04.

[116] Briz T, Ward RW. Consumer awareness of organic products in Spain: An application of multinominal logit models. Food Policy. 2009;34:295-304. DOI: 10.1016/j.foodpol. 2008.11.004.

[117] Olivas R, Díaz M, Bernabeu R. Structural equation modeling of lifestyles and consumer attitudes towards organic food by income: A Spanish case study. Ciencia e Investigación Agraria. 2013;40:265-277. DOI: 10.4067/S0718-16202013000200003.

[118] Mesías FJ, Escribano M, Gaspar P, Pulido F. Consumers' attitudes towards organic, PGI and conventional meats in extremadura (Spain). Archivos de Zootecnia. 2008;57:139-146.

[119] Mesías FJ, Martínez-Carrasco F, Martínez-Paz JM, Gaspar P. Willingness to pay for organic food in Spain: An approach to the analysis of regional differences. ITEA Información Técnica Económica Agraria. 2011;107:3-20. DOI: 10.1111/j. 1747-6593.2011.00286.x.

[120] Nunes B, Bennett D, Júnior SM. Sustainable agricultural production: An investigation in Brazilian semi-arid livestock farms. Journal of Cleaner Production. 2014;64:414-425. DOI: 10.1016/S0959-6526(00)00013-5.

[121] Adams DC, Salois MJ. Local versus organic: A turn in consumer preferences and willingness to pay. Renewable agriculture and Food Systems. 2010;25:331-341. DOI: 10.1017/S1742170510000219.

[122] Pugliese P, Zanasi C, Atallah O, Cosimo R. Investigating the interaction between organic and local foods in the Mediterranean: The Lebanese organic consumer's perspective. Food Policy. 2013;39:1-12. DOI: 10.1016/j.foodpol.2012.12.009.

[123] Rikkonen P, Kotro J, Koistinen L, Penttilä K, Kauriinoja H. Opportunities for local food suppliers to use locality as a competitive advantage - a mixed survey methods approach. Acta Agriculturae Scandinavica, Section B—Soil & Plant Science. 2013;63:29-37. DOI: 10.1080/09064710.2013.783620.

[124] Schleenbecker R, Hamm U. Consumers' perception of organic product characteristics. A review. 2013. Appetite. 2013;71:420-429. DOI: 10.1016/j.appet.2013.08.020.

[125] Campbell BL, Mhlanga S, Lesschaeve I. Perception versus reality: Canadian consumer views of local and organic. Canadian Journal of Agricultural Economics. 2013;61:531-558. DOI: 10.1111/j.1744-7976.2012.01267.x.

[126] Costanigro M, Kroll S, Thilmany D, Bunning M. Is it love for local/organic or hate for conventional? Asymmetric effects of information and taste on label preferences in an experimental auction. Food Quality and Preference. 2014;31:94-105. DOI: 10.1016/ j.foodqual.2013.08.008.

[127] Gracia A, Barreiro-Hurlé J, López-Galán B. Are Local and Organic Claims Complements or Substitutes? A Consumer Preferences Study for Eggs. Journal of Agricultural Economics. 2014;65:49-67. DOI: 10.1111/1477-9552.12036.

[128] Zoiopoulos P, Hadjigeorgiou I. Critical overview on organic legislation for animal production: towards conventionalization of the system? Sustainability. 2013;5:3077-3094. DOI: 10.3390/su5073077.

[129] A.J. Escribano, P. Gaspar, F.J. Mesías, M. Escribano, F. Pulido, (Ed. Victor R. Squires) 2015 Comparative Sustainability Assessment of Extensive Beef Cattle Farms in a High Nature Value Agroforestry System. In book: Rangeland Ecology, Management and Conservation Benefits, Publisher: Nova Publishers.

8

Quality and Nutrient Contents of Fruits Produced Under Organic Conditions

Taleb Rateb Abu-Zahra

Additional information is available at the end of the chapter

Abstract

Organic farming is an agricultural practice that raises plants especially vegetables and fruits without the use of synthetic pesticides, herbicides, fertilizers, or plant growth regulators. All over the world, the interest for organic farming has increased recently. Different greenhouse experiments were carried out in the northern Jordan Valley, to compare the effect of four fermented organic matter doses (1.5, 3.0, 4.5, and 6.0 kg m^{-2}), or different organic matter sources (cattle, poultry, and sheep manure in addition to 1:1:1 mixture of the three organic matter sources) with that of the conventional fertilizer and control treatments on different fruit quality parameters.

Results obtained showed that fruit titratable acidity (TA) percentage, size, moisture content, and ammonium and nitrate contents were higher in the conventionally produced fruits in comparison to the organically produced fruits. The organic treatments tended to produce fruits with higher anthocyanin, total soluble solids (TSS) percentage, dry matter content, ascorbic acid, total phenols, and crude fibre content in comparison to the control and conventionally produced fruits. In most cases, sheep manure source and 4.5 kg O.M m^{-2} treatment amount produced the best results.

Keywords: Nutrients, pigments, quality

1. Introduction

1.1. Environmental Issues

Environmental issues are capturing more and more of the world's attention; therefore, researchers and scientists are aiming at improving environmental quality through the adoption of techniques and measures that have a reduced impact on the environment [1]. Conventional agriculture practices utilize high-yield crop cultivars, chemical fertilizers and pesticides, irrigation techniques, and mechanization that have a huge impact on our environment [2]. Plants are subjected to attack by a large and diverse number of pathogens and pests; as a result, crop producers often use large amounts of agrochemicals in an attempt to improve and protect the fruit quality and plant vigor [3]. Ever since people have become aware that health is linked to health environment, the control and reduction of pollution have become the focus of worldwide concern [4]. Pollution is becoming a serious problem in agricultural regions; for example, various mineral fertilizers and agrochemicals lead to pollution and serious health problems in humans, hence alternative production techniques which employ biological or organic compounds for disease and pest control are needed [5]. In addition to the human health concern of elevated heavy metal concentrations in soil, they can cause harm to native ecosystem and accumulation in plant tissue can result in damage to wildlife [6]. Plant toxicity is the primary concern for elevated zinc concentration in soil, whereas the potential for risk to the herbivores is the primary concern with elevated cadmium concentration in soil, while human health concerns focus on lead concentration for which the most pertinent pathway is direct ingestion of soil [7].

1.2. Organic culture

Organic farming, which essentially excludes the use of many inputs associated with modern farming, most notably synthetic pesticides and fertilizers, is becoming more and more popular worldwide [2, 8]. Consumer's awareness of the relationship between foods and health, together with environment concerns, has led to an increased demand for organically produced foods. In general, the public perceives organic foods as being healthier and safer than those produced through conventional agricultural practices [9]. Consumers demand organic products because they believe they are more favorable and respectful to the environment and human health [10]. Organic foods have a nutritional and sensory advantage in comparison to their conventionally produced counterparts. Advocates for organic produce claim that it contains fewer harmful chemicals, is better for the environment, and may be more nutritious [11].

2. Fruit nutrient contents

2.1. Mineral contents

Mineral contents of fruits were found to be higher in fruits produced under conventional systems in comparison to the fruits produced under organic systems [12]. For example, bell pepper fruits, which were produced under conventional systems, were characterized by a high content of minerals (Table 1). The highest contents of zinc and iron in bell pepper were obtained in the conventional treatment with significant differences between other treatments, while there were no significant differences among the organic matter treatments, which could be attributed to the high application of chemical fertilizers [13].

Treatments	Zinc content (ppm)	Iron content (ppm)
Conventional	1.410 a	57.75 a
Cattle manure	1.170 b	45.50 b
Poultry manure	1.163 b	39.75 c
Sheep manure	1.165 b	39.25 c
Mixture manure	1.227 b	42.75 bc

*Means within each column having different letters are significantly different according to Least Significant Difference at 5% level.

Table 1. Effect of culture systems on contents of zinc and iron in bell pepper fruit

The contents of calcium, magnesium, sodium, potassium, and phosphorous in bell pepper fruit were significantly higher in those produced with conventional system than all those produced with organic matter systems (Table 2); even though the highest calcium content was obtained by the conventional treatment, there was no significant difference with the poultry manure, which could be due to the high use of limestone in the chicken food mixture [13].

Treatments	Calcium (mg 100 g^{-1})	Magnesium (mg 100 g^{-1})	Phosphorus (mg 100 g^{-1})	Sodium (mg 100 g^{-1})	Potassium (mg 100 g^{-1})
Conventional	260 a	89.25 a	394 a	26.1 a	2323 a
Cattle manure	243 b	79.50 b	315 b	19.1 b	1889 bc
Poultry manure	257 a	81.75 ab	362 ab	19.9 b	1820 c
Sheep manure	239 b	84.50 ab	349 ab	18.1 b	1986 b
Mixture manure	246 b	77.75 b	348 ab	19.6 b	1915 bc

*Means within each column having different letters are significantly different according to Least Significant Difference at 5% level.

Table 2. Effect of culture systems on contents of calcium, magnesium, phosphorus, sodium, and potassium in bell pepper fruit

2.2. Ammonium and nitrate

Vegetables represent the most important source of nitrogen for human nutrition, which is essential for growth. Therefore, its accumulation in plants is a natural phenomenon resulting from uptake of the nitrate ion that is found in excess amounts, and the intensive use of nitrogen fertilizer and manure causes nitrate contamination of the environment; therefore, vegetables can accumulate high levels of nitrogen and, when consumed, pose serious health concerns [13]. Ammonium and nitrate contents in conventionally grown strawberry fruits were 49.4 and 23.6 ppm, respectively, due to high use of inorganic nitrogen fertilizers, whereas it was found that ammonium content was 32.3 ppm and nitrate content was extremely low in organically produced fruits [10]. The nitrate content in bell pepper fruit was very low (<200 mgkg^{-1}), for all different cultural systems (organic or inorganic), even though the minimum value of nitrate content for organically produced bell peppers and the maximum value for fertilized bell peppers were found below the safe limit [13].

3. Fruit quality

3.1. Total soluble solids and titratable acidity

All organically produced fruits had significantly higher total soluble solids (TSS) and lower titratable acidity (TA) in comparison to the conventionally produced fruits [5, 14]; for example, sensory attributes are important aspects of fruit quality, and the balance between sweetness and sourness are the most important determinants of overall quality of fruits [15]; for example, acceptance of the flavor quality of strawberry fruits is minimum 7% for TSS content, while the maximum is 0.8% for TA [16]. Organically grown strawberries had significantly higher TSS (7.1%) and lower TA content (0.93%) in comparison to the conventionally grown strawberries that had 6.6% TSS and 0.99% TA. On the other hand, addition of animal manure improved bell pepper fruit taste by increasing the percentage of TSS and the addition of animal manure decreased the percentage of TA in bell pepper fruit [10].

3.2. Total phenols

Phenolic metabolites may suit human health and contribute to the prevention of chronic diseases such as cancer and cardiovascular diseases [17]. In addition, phenolic compounds play a vital role in plant defense mechanisms against insect, fungi, and animal herbivores [18]. Levels of phenolic compounds were higher in organically grown fruits than the levels in conventionally produced fruits, because the restricted use of herbicides, pesticides, insecticides, and chemical fertilizers was reported to accelerate synthesis of phenolic compounds in organically produced fruits [19].

3.3. Ascorbic acid (Vitamin C)

Ascorbic acid content in fruits is cultivar dependent according to Leskinen et al. [20]; levels of ascorbic acid in organically produced fruits were consistently higher than the levels in the conventionally grown ones [8]. The highest fruit ascorbic acid content (50.5 mg 100 g^{-1} fruit fresh weight) was obtained by the organically treated berry fruits, whereas the conventional treatment gave the lowest ascorbic acid content (41.25 mg 100 g^{-1} g fruit fresh weight), according to Abu-Zahra et al. [10]. On the other hand, Cayuela et al. [14] did not find significant difference in the ascorbic acid content between organic and conventional grown strawberry fruits. Also manure type has an effect; the highest amount of vitamin C was obtained from the sheep manure–treated pepper fruits, while the lowest amount was obtained by the conventionally produced pepper fruits [10].

3.4. Crude fiber

Fruit crude fiber content highly differs according to fruit dry weight [21], but it is found to be higher in organically produced fruits in comparison to conventionally produced fruits [10]; the high crude fiber content in the organically produced fruits could ensure better nutritional and health benefits related to fiber consumption [22]. The highest strawberry crude fibre fruit value (8.13%) was obtained by the 4.5 kg organic matter/m^2, which was significantly different from the conventional, and control treatments [13]. Although, crude fiber of bell pepper fruit was improved by the use of the cattle manure which produced the highest (2.96%) crude fiber content in comparison to the conventional system which produced the lowest content (2.8%) [23].

3.5. Fruit size

Fruit size is highly affected by the farming systems; the conventional agriculture resulted in the biggest fruits, in comparison to organically produced fruits. The large fruit size in the conventional farming system may be due to the good availability of soil nutrients that produced vigorous plants with higher yield and larger fruits. But it was observed that the use of high amount of organic matter (6 kg O.M/m^2) produced a large fruit size, which may be due to the good improvement of physical and chemical properties of the soil [10, 24].

3.6. Fruit fresh weight

Fruit weight depends on cultivar and temperature rather than on the culture system (organic or conventional) [10]. Moreover, most researchers found only small and non-significant differences between organic and conventional systems in respect to fruit weight [20]. But in an experiment conducted on strawberry plants, they observed that the use of chemical fertilizers were found to produce the highest significant average fruit weight compared to fruits produced by using organic materials or without using any type of fertilizers [10, 25].

3.7. Fruit moisture content and dry weight

Fruit moisture content showed an opposite trend to fruit dry matter content; organically produced fruits had more dry matter and lower water content in comparison to the conventionally produced ones. The decrease in fruit water content of the organically produced fruits was reflected on increasing fruit dry matter content in comparison to the conventionally produced fruits that produced the lowest fruit dry matter and highest water content [10]. For example, the highest strawberry moisture content (93.37%) was obtained by the conventional system which produced the lowest fruit dry matter content (6.63%), while strawberry fruits that are produced under organic systems, contains 92.61% moisture content and 7.39% of dry matter content [10].

3.8. Fruit pH

The fruit taste is highly affected by the fruit pH; addition of organic materials was found to lower the strawberry fruit pH, especially by using sheep manure as a source of organic matter

[24]. However, in an experiment conducted on pepper plant, results do not show any significant differences between all of the used organic and inorganic treatments on fruit pH [23].

4. Fruit pigments

4.1. Chlorophyll

Chlorophyll content of the leaves was increased by the use of organic matter applications; the highest increase was obtained by using the sheep manure as a source of organic matter, while the lowest amounts of leaf chlorophyll content were obtained by the use of chemical fertilizers [26].

A promotional effect of organic matter treatments on chlorophyll contents might be attributed to the fact that nitrogen is a constituent of chlorophyll molecule [3]; moreover, nitrogen is the main constituent of all amino acids in protein and lipids that act as a structural compound of the chloroplast. Contradictory data about the relationship between growth and chlorophyll content of leaves have been reported in which bio-fertilizers increased the content of photosynthetic pigments [27].

4.2. Anthocyanin

Organically grown fruits developed a significantly stronger color than conventionally grown ones [14]. The highest anthocyanin content of strawberry fruits (42.88 mg 100 g^{-1}fruit fresh weight) was obtained by the 6 kg O.M/m^2 treatment, while the least anthocyanin content was obtained by the control treatment (neither synthetic fertilizers nor organic materials). In spite of that, the anthocyanin content of the control treatment of strawberry plants remained within the ranges between 17.8 and 41.8 mg 100 g^{-1}, and values lower or higher than that range should not be acceptable [10].

In another study conducted on red pepper fruits, the highest anthocyanin (38.5 mg 100 g^{-1}) amount was obtained by the mixture of different organic matter treatment. And the least anthocyanic content was obtained by the conventional culture system, which proves that organic farming provides peppers with the highest intensities of red and yellow colors, while the conventional fruits were those with the lowest values of color intensity [23].

4.3. Lycopene

It is recorded that fruit lycopene content was the highest in conventional agriculture, but without significant differences from the different organic matter sources. Also fruit lycopene was affected by the organic matter source, and the lowest lycopene content was obtained by the poultry manure source–treated pepper fruits, which means lycopene fruit content does not improve by the use of organic matter treatments in comparison to conventional agriculture that hastened fruit lycopene content [23].

5. Conclusions

Fruit characteristics from plants cultivated in soil supplemented with animal manure were generally better than those from plants grown in soils only or supplemented with chemical fertilizers. In most cases of animal manure sources, sheep manure gave the best results. On the other hand, the use of chemical fertilizers was found to increase the fruit lycopene content and improve fruit size and yield by increasing the fruit weight. Organic foods contain fewer harmful chemicals, are better for the environment, and may be more nutritious.

Author details

Taleb Rateb Abu-Zahra

Address all correspondence to: talebabu@yahoo.com

Department of Plant Production and Protection, Faculty of Agricultural Technology, Al-Balqa Applied University, As-Salt, Jordan

References

[1] Hamdar, B. C., and Rubeiz, I. G. 2000. Organic farming: Economic efficiency approach of applying layer litter rates to greenhouse grown strawberries and lettuce. Small Fruits Review. 1(1): 3-14.

[2] Ames, G., Born, H., and Guerena, M. 2003. Strawberries: Organic and IPM options. NCAT agriculture specialists, ATTRA. Retrieved from http://attra.ncat.org/attra-pub/PDF/strawberry.pdf (access 2008).

[3] Abu-Zahra, T. R. 2012. Vegetative, flowering and yield of sweet pepper as influenced by agricultural practices. Middle-East Journal of Scientific Research. 11(9): 1220-1225.

[4] Vasile, G., Artimon, M., Halmajan, H., and Pele, M. 2010. Survey of Nitrogen Pollutants in Horticultural Products and Their Toxic Implications. Proceeding of the International Conference Bioatlas, Transylvania, University of Brasov, Romania.

[5] Turemis, N. 2002. The effects of different organic deposits on yield and quality of strawberry cultivar Dorit (216). Acta Horticulturae. 567: 507-510.

[6] Beyer, W. N. 2000. Hazards to wildlife from soil-borne cadmium reconsidered. Journal of Environmental Quality. 29:1380-1384.

[7] Brown, S., Chaney, R., Hallfrisch, J., Rayan, J. A., and Berti, W. R. 2004. *In situ* soil treatments to reduce the phyto- and bioavailability of lead, zinc, and cadmium. Journal of Environmental Quality. 33: 522-531.

[8] Asami, D. K., Hong, Y. J., Barrett, D. M., and Mitchell, A. E. 2003. Comparison of the total phenolic and ascorbic acid content of freeze-dried and air-dried marionberry, strawberry, and corn grown using conventional, organic, and sustainable agriculture practices. Journal of Agricultural and Food Chemistry. 51: 1237-1241.

[9] Jolly, D. A. 1989. Organic foods-consumer attitudes and use. Food Technology. 43(11): 60.

[10] Abu-Zahra, T. R., Al-Ismail, K., and Shatat, F. 2007. Effect of organic and conventional systems on fruit quality of strawberry (*Fragaria* X *Ananassa* Duch) grown under plastic house conditions in the Jordan Valley. Acta Horticulturae. 741: 159-172.

[11] Mitchell, A. E., and Chassy, A. W. 2005. Antioxidants and the nutritional quality of organic agriculture. Retrieved from http://mitchell.ucdavis.edu/Is%20Organic %20Better.pdf (access 2006)

[12] Jadczak, D., Grzeszuczuk, M., and Kosecka, D. 2010. Quality characteristics and content of mineral compounds in fruit of some cultivars of sweet pepper (*Capsicum annum* L.). The Elemental Journal. 15(3): 509-515.

[13] Abu-Zahra, T. R., Ta'any, R. A., Tahboub, A. B., and Abu-Baker, S. M. 2013. Influence of agricultural practices on soil properties and fruit nutrient contents of bell pepper. Biosciences Biotechnology Research Asia. 10(2): 489-498.

[14] Cayuela, J. A., Vidueira, J. M., Albi, M. A., and Gutierrez, F. 1997. Influence of the ecological cultivation of strawberries (*Fragaria* X *Ananassa* Cv. Chandler) on the quality of the fruit and on their capacity for conservation. Journal of Agricultural and Food Chemistry. 45: 1736-1740.

[15] Shamaila, M., Baumann, T. E., Eaton, G. W., Powrie, W. D., and Skura, B. J. 1992. Quality attributes of strawberry cultivars grown in British Columbia. Journal of Food Science. 57: 696-699.

[16] Kader, A. A. 1999. Fruit maturity, ripening, and quality relationships. Acta Horticulturae. 485: 203-208.

[17] Torronen, R., and Maatta, K. 2002. Bioactive substances and health benefits of strawberries. Acta Horticulturae. 567: 797-803.

[18] Cheng, G. W., and Breen, P. J. 1991. Activity of phenylalanine ammonia-lyase (PAL) and concentration of anthocyanins and phenolics in developing strawberry fruit. Journal of American Society for Horticultural Science. 116: 865-869.

[19] Hakkinen, S. H., and Torronen, A. R. 2000. Content of flavonols and selected phenolic acids in strawberries and *Vaccinium* species: Influence of cultivar, cultivation site and technique. Food Research International. 33: 517-524.

[20] Leskinen, M., Vaisanen, H. M., and Vestergaard, J. 2002. Chemical and sensory quality of strawberry cultivars used in organic cultivation. Acta Horticulturae. 567: 523-526.

[21] Pellet, P. L., and Shadarevian, S. 1970. Food Composition: Tables For Use in Middle East (2nd ed.). American University of Beirut, Lebanon.

[22] Anderson, J. W., Smith, B. M., and Gustafson, N. J. 1994. Health benefits and practical aspects of high-fibre diets. American Journal of Clinical Nutrition. 59: 1242-1247.

[23] Abu-Zahra, T. R. 2011. Influence of agricultural practices on fruit quality of bell pepper. Pakistan Journal of Biological Sciences. 14(18): 876-881.

[24] Abu-Zahra, T. R., and Tahboub, A. A. 2009. Strawberry (*Fragaria* X *Ananassa* Duch) fruit quality grown under different organic matter sources in a plastic house at Humrat Al-Sahen. Acta Horticulturae. 807: 353-358.

[25] Abu-Zahra, T. R., and Tahboub, A. A. 2008. Strawberry (*Fragaria* X *Ananassa* Duch) growth, flowering and yielding as affected by different organic matter sources. International Journal of Botany. 4(4): 481-485.

[26] Tahboub, A. A., Abu-Zahra, T. R., and Al-Abbadi, A. A. 2010. Chemical composition of lettuce (*Lactuca sativa*) grown in soils amended with different sources of animal manure to stimulate organic farming conditions. Journal of Food, Agriculture & Environment. 8(3 & 4): 736-740.

[27] Arisha, H. M., and Bradisi, A. 1999. Effect of mineral fertilizers and organic fertilizers on growth, yield and quality of potato under sandy soil conditions. Zagazig. Journal of Agricultural Research. 26: 391-405.

PERMISSIONS

LIST OF CONTRIBUTORS

Alfredo Aires
Centre for the Research and Technology of Agro-Environment and Biological Sciences (CITAB), Universidade de Trás-os-Montes e Alto Douro, UTAD, Quinta de Prados, Portugal

Mehdi Zahaf
Telfer School of Management, University of Ottawa, Canada

Madiha Ferjani
Mediterranean School of Business, Tunisia

Elpiniki Skoufogianni and Konstantinos Martinos
Laboratory of Agronomy and Applied Crop Physiology, Dept. of Agriculture, Crop Production and Rural Environment, University of Thessaly, Volos, Greece

Alexandra Solomou
National Agricultural Research Foundation, Institute of Mediterranean Forest Ecosystems Terma Alkmanos, Ilisia, Athens, Greece

Aikaterini Molla
National Agricultural Research Foundation, Larisa, Greece

Jan Moudry Jr, Jan Moudry and Zuzana Jelinkova
University of South Bohemia in Ceske Budejovice, Czech Republic

Maliha Sarfraz
Institute of Pharmacy, Physiology and Pharmacology, University of Agriculture Faisalabad, Pakistan

Mushtaq Ahmad
Department of Plant Sciences, Quaid-i-Azam University Islamabad, Pakistan

Wan Syaidatul Aqma Wan Mohd Noor
School of Biosciences and Biotechnology, Faculty of Science and Technology, University Kebangsaan Malaysia (UKM) Bangi, Selangor, Malaysia

Muhammad Aqeel Ashraf
Department of Environmental Science and Engineering, School of Environmental Studies, China University of Geosciences, Wuhan, P. R. China
Water Research Unit, Faculty of Science and Natural Resources, University Malaysia Sabah, Kota Kinabalu, Sabah, Malaysia

Florentina Sauca and Catalin Lazar
National Agricultural Research and Development Institute (NARDI) – Fundulea, Calaraşi, Romania

Alfredo J. Escribano
Researcher and consultant. C/ Rafael Alberti, Cáceres, Spain

Taleb Rateb Abu-Zahra
Department of Plant Production and Protection, Faculty of Agricultural Technology, Al-Balqa Applied University, As-Salt, Jordan

Index